ENGLISH HISTORIC CARPENTRY

ENGLISH HISTORIC CARPENTRY

by

Cecil A.Hewett

Linden Publishing
2006 South Mary Street
Fresno, California 93721
1-800-345-4447

Front Cover: GHL6199 Guildhall Library and Museum of the Corporation of London, lithograph by Canton, 1872. Guildhall Library, Corporation of London/Bridgeman Art Library, London.

ENGLISH HISTORIC CARPENTRY
by
Cecil A. Hewett

123456789

ISBN: 0-941936-41-4

Library of Congress Cataloging- in -Publication Data

Hewett, Cecil Alec.
  English historic carpentry / by Cecil A. Hewett. -- 1st Linden ed.
    p. cm.
  Reprint. Originally pub. : London : Phillimore, 1980
  includes bibliographical references and index.
  ISBN 0-941936-41-4
  1. Carpentry--History. 2. Building, Wooden--England--History.
  3. Historic buildings--England--Dating. 4. England--Buildings,
  structures, etc. I. Title.
  TH5604.H47 1997
  694' .0942--dc21   97-11126
                                                                CIP

Originally published by
Phillimore & Co. Ltd.
Sussex, England
1980

First Linden Edition 1997
@ Cecil A. Hewett
Printed in the United States of America

Contents

List of Plates

(between pages 82 and 83)

The Illustrations

All the line drawings reproduced are by the author; their quality varies considerably because the dates of execution cover three decades. Those illustrating the D'Arcy Chantry, at Maldon, are by my colleague, Mr. M. C. Wadhams, to whom I am indebted for numerous discussions of the subject. Figs. 225 to 233, and 236 to 249 are reproduced with the kind consent of the Greater London Council, and that authority's Historic Buildings Section. The photographers whose work constitutes the half-tone plates are individually named in the List of Plates, and to them I am most grateful.

C.A.H.

Glossary

Addorsed: From heraldry, meaning placed back to back—the opposite of 'addressed', or face to face.

Angle tie: Tying timbers placed across angles, normally the returns of wall plates. These were widely used during the 18th century as a means to step hip rafters, which were seated in a third timber, the dragon piece.

Abutment: *Abut*, O.Fr. *abuter*, to touch at the end (*à*, to, *bout*, end). Any point in timber jointing where one timber's end touches another constitutes an abutment. A 'butt-joint' is, therefore, one where ends meet; no integration is implied.

Arcature: The curvature of an arch, as segmental, ogee or lancet.

Arcade: A range of arches. Term applied also to the series of posts standing inside an aisled timber building, because they are sometimes arch-braced in their longitudinal direction.

Arch braces: Term generally applied to braces beneath tie beams, which were frequently curved, or arched.

Arris: The edge at which two surfaces meet.

Arris-trenched: Trenched (q.v.), so that the trench is cut obliquely through an arris and affects both adjacent surfaces.

Ashlar pieces: Short, vertical timbers at the feet of rafters, generally standing upon sole pieces. These continue the internal wall surface until it meets the underside of the rafters, avoiding a visual discontinuity and greatly strengthening the rafters' base.

Barefaced: With the face uncovered, without a mask; avowed, open. Term used to denote a timber joint possessed of only one shoulder, but which normally possesses two.

Base crucks: Timbers placed as wall posts and containing the naturally grown angle of the eaves, above which they may rise to collar height.

Bays: The divisions, normally postulated by the material used for construction of the lengths of buildings. In the case of arcades, each arch is taken as one bay.

Bird's-mouth: Term used to describe joints bearing a visual resemblance to an open bird's beak.

Blade, -ing, -ed: Term used to specify scarfs that are face-halved and terminated in inset, barefaced tongues. See Fig. 270.

Bole: The butt of a tree trunk, normally of concave conoid form, used to provide jowls by inverting the timber.

Bowtell: Small roll moulding, or bead.

Brace: Any timber reinforcing an angle, usually subjected to compression.

Bridle: Term applied to timber joints having open-ended mortises, and tenons resembling a horse's mouth with the bit of the bridle in place.

Bressummer: Breast-summer, a timber extending for the length of a timber building, normally forming the sill of a jettied storey.

Bridging-joist: Floor timber that supports the ends of common joists, and normally bridges the bays from one binding-joist to the next.

Broach: A spit or point.

Butment cheeks: The timber left on either side of mortises, against which the shoulders of tenons abut.

Cant post: Posts that converge upwards; see Navestock belfry.

Camber beams: Beams sufficiently cambered to form the basis of the simplest type of roof, their curvature serving to drain their surfaces when clad.

Chamfer: The slope or bevel created by removing a timber's corner or arris. These are termed 'through' if run off the end of the workpiece, 'stopped' if terminating in a decorative form before the end of the piece or its conjunction with another.

Cant: The oblique line or surface which cuts off the corner of a square or cube. The term is applied to soulaces in roofs, because they produce a canted plane; roofs possessing soulaces, collars and ashlar pieces are thus described as 'of seven cants'.

Chase: From *chasse*, a shrine for relics. In carpentry a score cut length-wise, a lengthened hollow, groove, or furrow.

Chase, or chased mortise: A long mortise into which a tenon may be inserted sidewise.

Cladding: The external covering applied to a wall or a roof.

Clamp: A term applied variously to timbers depending upon the type of building. In houses the term denotes horizontal timbers attached to the wall studs in order to support floors; these clamps normally indicate the later intrusion of such floors.

Clench, clinch: Either to turn the point of a nail or spike, and re-drive it back into the timber through which it has passed; or to form its end into a rivet, or clench, by beating it out upon a washer or rove.

Coak: A peg or dowel of a diameter almost equal to its length, used in 19th-century shipwrighting to join futtocks and timbers because it was cheaper than scarfing them.

Cogging: A method of housing an entire timber's end, sometimes used to prevent its rotation—as in door cases.

Collar beam: A roof timber, placed horizontally and uniting a rafter couple at a point between the bases and the apex. Collar beams can act either as ties or strainers.

Collar purlin: A longitudinal timber uniting and supporting the collars; it is normally carried by crown posts.

Compass timber: A term denoting timber of natural and grown curvature, as distinct from relatively straight-grown timber from which curves are cut.

Common joist, -rafter: The majority of either kind, and normally those of the least cross section in any floor or roof.

Corner post: The post standing at the return of two walls, as at the end and adjacent side of a building.

Counter rebate: See Fig. 57.

Crenellate: To furnish with battlements, a decorative device used much in timber buildings.

Crown post: A vertical timber standing at the centre of a tie beam and supporting a collar purlin.

Cruck blade: The elbowed timber forming one half of a pair of crucks.

Cusp: In Gothic tracery the pointed shape or form created by the intersection of two concave arcs.

Cyma: Ogee. Formed by a concave and a convex arc in a single linear association.

Dormer: An upright window protruding from the pitch of a roof.

Double tenons: Two tenons cut from the same timber's end and placed in line; if side by side they constitute a pair of single tenons.

Dragon beam, -piece: A timber bisecting the angle formed by two wall plates. If a beam it supports a jetty continued around the angle; if a piece it normally serves to step a hip rafter.

Draw knife, or -shave: A hand tool possessing handles at either end of its blade, used to produce chamfers.

Durns: Timbers with grown bends suitable for the manufacture of door-ways of Gothic arcatures; two were frequently sawn from one piece of the requisite form.

Eaves: The underside of a roof's pitch that projects outside a wall.

End girt, -girth: Horizontal timber in an end wall placed halfway betwixt top plate and groundsill, thereby shortening the studs and stiffening the wall.

Fascia: A board forming a front, frequently used to cover a number of timbers' ends, as joists at a jetty, or rafters at an eaves.

Fillet: In moulding profiles a small raised band, normally of square section. Also a small squared timber.

Fish: A length of timber with tapering ends which can be used to cover and strengthen a break in another timber.

Fished scarf: A scarf that relies upon the introduction of a third timber.

Footing: Foundation.

Free tenon: A tenon used as a separate item, both ends being fitted into mortises cut into two timbers to be joined; often used to effect edge-to-edge joints.

Gambrel: Perhaps Old French (Norman). 'Also gambrel roof . . . so called from its resemblance to the shape of a horse's hind leg' (*O.E.D.*, 1933).

Girth, girt: Horizontal timbers in wall frames, placed at half height, which shorten and thereby stiffen the studs.

Groundsill: The first horizontal timber laid for a timber building. As the name implies these were in ancient times laid directly upon the ground, as was the case at Greensted church, Essex.

Halving: In jointing the removal of half the thicknesses of two timbers, as in cross halving.

Harr, arr: The edge timber of a door leaf or gate nearest the hinges, the opposite edge timber of which is the head.

Hanging knee: Term denoting a knee placed beneath a beam. Knees placed above beams are 'standing knees, or standards', and those in the horizontal plane are 'lodging knees'. All three derive from shipwrighting and were used in ships as early as the Viking period.

Haunch: Adjuncts of tenons, designed to resist winding; they may be square or diminished.

Header: Short timber to carry rafters' tops at the exiture of chimney-stacks.

Hewn knee: A knee, or angle, cut from a timber's end, as distinct from a separate and applied piece.

Hip rafter: A rafter pitched on the line of intersection of two inclined planes of roof, forming the arris of a pyramidal form.

Hogging: Stress caused by supporting the centre of a beam and leaving the ends unsupported, as when a wave rises amidships of a vessel, and beneath her.

Housing: In jointing a cavity large enough to hold an entire timber's end.

Jamb: The side of a doorway, archway or window.

Jetty: The projection of a floor outside its substructure, upon which the next storey was built. This resulted in floor areas that increased as the storeys ascended.

Joggle: Said of two timbers, both of which enter a third but which have their joints out of line in order to avoid excessive weakening.

Joist: The horizontal timbers supporting floors; these are binding, bridging, common and trimming. Binders unite storey posts; bridgers span or bridge each bay from binder to binder; commons are the most numerous and actually carry the floor boards. Trimmers are used to frame the edges of voids, such as stair wells.

Jowl: (Jole) 'The external throat or neck when fat or prominent . . . the dewlap of cattle'. (*O.E.D.*) Term applied to the thickened ends of such timbers as storey posts which facilitate the jointing of several other timbers.

Kerf: The cut produced by a saw.

Key: Tapered piece of dense hardwood transfixing a scarf, used to close its abutments.

King stud: A stud placed centrally in a gable, normally supporting the collar purlin.

Lap dovetail: That form of dovetail that overlaps, and is not finished flush. The alternative form is the 'through dovetail' used by cabinet makers; they may also be 'secret', 'secret-mitred' or double.

Lap joints: Any jointed timbers which overlap each other.

Lodged: 'To put and cause to remain in a specified place.' (*O.E.D.*) A term applied to floors retained in place by their weight alone.

Main span: In aisled buildings this is the central and greatest distance spanned.

Midstrey: The porch-like structure at the front of a barn, derived from middle-strey. Each bay of a barn was a strey, and ancient barns normally had one such porch at their centre.

Mitre: Abutments at 45 degrees, producing square returns.

Muntins: Vertical members of panelled areas; the term may derive from mountants.

Mullions: Vertical components of windows, placed in the void.

Nogging: The material used to infill a framed wall betwixt sill and top.

Notched laps: A category of lap joints having V-shaped indentations on plan-view to prevent their lengthwise withdrawal.

Outshot, outshut: An area of space added to a building's bays, normally at the sides: when at the end of a building they are called by the ancient term 'culatia'.

Passing brace: A brace uniting several successive members of a frame and passing them by means of halved jointing; of mainly Early English and Decorated usage. (Author's coinage, 1962, without historical validity.)

Plate: A horizontal timber laid at the base of a timber frame; the term implies a footing, as distinct from a groundsill.

Prebend: The share of the revenues of a cathedral or collegiate church allowed to a clergyman who officiates in it at stated times.

Prick post: Any vertical timber placed in compression, but not a storey post.

Principal-rafter: A heavy rafter placed at bay intervals, normally associated with side purlins.

Purlin: A longitudinal timber in a roof.

Queen posts: Posts set in pairs between tie beams and collars and acting in compression.

Raking struts: Inclined struts used in pairs between tie beams and principal-rafters.

Reversed-assembly: Indicates a system of rearing transverse framing units, the lengthwise timbers of which (top plates) are laid last. In these cases the tie beams are *under* the top plates. (Author's coinage, 1962, without historical validity.)

Rive: To split timber lengthwise, i.e. along its grain.

Rove: The circular plate, or washer, upon which the clench, or rivet, in boat- or ship-building is formed.

Sagging: Stress caused by supporting the ends of a timber and applying weight to its centre.

Sally: An obtusely angular and pointed projection, normally on a timber's end. Alternatively a 'tace'.

Samson post: 'Pillar erected in a ship's hold, between the lower deck and the Kelson'. (*O.E.D.*) The term alludes to the strength of Samson (Judges XVI, 29), and is applied to similar posts used to support early floors.

Scarfing: The jointing of relatively short timbers into continuous lengths, by means of various expedients; the four faces are smooth and continuous.

Scarfed cruck: A cruck blade having a scarf-jointed angle, as distinct from a grown angle.

Set: The divergence of the sides of a dovetail.

Shore: An inclined timber supporting a vertical one, acting in compression.

Shuts: The edges of a door leaf collectively form the 'shuts' of that door.

Side girt, girth: See end girt.

Soffit: Underside or archivolt.

Sole pieces: Short horizontal timbers forming the base of any raftering system that has a base triangulation.

Soulace: A definitive term (Salzman, 1952) for secondary timbers connecting rafters with collars, and placed *under* the latter.

Spur tie: A short tie such as connects a cruck blade and a wall plate, or a collar arch and a wall plate.

Spire mast: Central vertical timber of a framed spire.

Squint: Angle other than 90 degrees.

Storey post: A wall post of a multi-storeyed timber building that continues through the floor levels.

Straining beam: A horizontal beam between two posts, acting in compression to keep them apart.

Strut: A timber in a roof system that acts in compression, in a secondary capacity.

Stub tenon: A short tenon that does not entirely penetrate the mortised concomitant timber.

Studs: From O.E. *studu*, a post; the vertical common timbers of framed timber walls.

Table: A raised rectangular portion on a worked timber, normally a scarf adjunct.

Tace: See sally.

Tail: The male part of a dovetail joint.

Tie beams: Beams laid across buildings to tie both walls together; they must have unwithdrawable end joints for this purpose.

Tongue: A fillet worked along the edge of a plank to enter a groove in another.

Top plate: A horizontal timber along the top of a framed wall.

Transom: A cross beam acting as a support for the superstructure.

Trench: A square sectioned groove cut across the grain.

Tusk: The wooden key driven through the protruding end of a tusked tenon, an unwithdrawable form of that joint.

Waney: Used to describe timber, the squared section of which is the greatest that can be cut from the rounded trunk, when any missing sharp arrises are said to be 'waney' edges.

Winding: In carpentry the result of torque or twisting, or the result of drying a spirally-grained tree.

Wind braces: Braces fitted into the angles of either roofs or walls to resist wind pressures.

Only terms that are likely to be strange to the general reader are here given, and the correct reference is, of course, the *Oxford English Dictionary*.

Acknowledgments

I wish to record my gratitude and indebtedness to all who have facilitated my studies during the past 30 years: the bishops, deans and chapter clerks, master masons and masons, carpenters and librarians of our great cathedrals—and the numerous architects responsible for their fabrics. Also to the many vicars and rectors of parish churches, and their secretaries to Parochial Church Councils. Among these I particularly thank Mr. L. S. Colchester of Wells Cathedral, for minutely detailed information concerning the documentation of Wells; Dr. J. H. Harvey for his unfailing readiness to assist with all matters that affect the dating of the cathedrals or monastic great churches of England; and Mr. R. O. C. Spring of Salisbury Cathedral, for his sustained kindness and assistance with studies and ascents of that cathedral. Also Dr. H. M. Taylor, for much advice and assistance with the assessment of pre-Conquest buildings. I am also grateful to the late Mr. S. E. Rigold for much assistance with documentary researches concerning many manor-houses, and parts of the historical background information for these I quote from his writings. I am particularly indebted to Miss Linden Lawson of Phillimore for her immense care and attention to detail throughout the period of this book's production.

I thank the innumerable owners and occupiers of countless historic buildings, both domestic and agricultural, who have so kindly allowed their roofs to be searched, or their floors excavated in search of early bases for posts. Also my wife, for continual assistance and tolerance during the long and irksome compilation of this and previous books upon the subject.

Bibliography

Arnold, T., *Memorials of St. Edmund's Abbey*, Rolls series, 96 (London, 1890).

Barker, F., ed. and trans., *Vita Aedwardi Regis* (London, 1890).

Beaumont, G. F., *A History of Coggeshall in Essex* (London, 1890).

Berger, R., ed., *Scientific Contributions to Medieval Archaeology* (1970).

Bilson, J., 'Notes on the earlier Architectural History of Wells Cathedral', *Archaeological Journal* (1928).

Brown, G. Baldwin, *The Arts in Early England II, Anglo-Saxon Architecture* (London, 1903. 2nd ed. 1925).

Bruce-Mitford, R. L. S., *The Sutton Hoo Ship Burial* (B.M. Trustees, London, 1954).

The Builder, Vol. LXXXVII (1904).

Calendar of Patent Rolls, Vol. II, Edward VI (H.M.S.O.).

Cave, P., *Hospital of St. Cross* (1970).

Chapman, F. R., ed., *The Sacrist Rolls of Ely*, Vol. I (Cambridge, 1907).

Chisenhall-Marsh, T. C., *Domesday Book, relating to Essex* (Chelmsford, 1864).

Christensen, A. E., jnr., *Vikingskipene* (Oslo, 1970).

Chubb, Rev. N., *All Saints, Brixworth* (1977).

Clapham, A. W., *English Romanesque Architecture, After the Conquest* (1964).

Clowes, G. S. Laird, *Sailing Ships, their History and Development* (H.M.S.O., 1932).

Colchester, L. S. and Harvey, Dr. J. H., 'Wells Cathedral', *Arch. Jnl.*, Vol. 131 (1974).

Colchester, L. S., *Wells Cathedral Library* (1978) 2nd ed.

Colvin, H. M., ed., *History of the King's Works* (1963).

Cristie, H., Olsen, O. and Taylor H. M., 'The Wooden Church of St. Andrew at Greensted, Essex', *Antiquaries Journal* (1979), Vol. LIX, part I.

Davey, Dr. N., *A History of Building Materials* (London, 1961).

Davis, R. H. C., 'The Norman Conquest', *Hist. Ass. Jnl.*, Vol. LI (1966).

Deneux, H., 'L'Evolution des Charpentes du xie au xve siècle' (C.R.M.H.).

Downes, K., *Hawksmoor* (1969).

Drinkwater, N., 'Old Deanery at Salisbury', *Antiquaries Journal*, 44 (1964).

Dugdale, W., *Monasticon Anglicanum* (London, 1718).

Everett, C. R., 'Notes on the Decanal and other houses in the Close of Sarum', *Wiltshire Archaeological & Natural History Magazine*, 50 (1944).

Finn, R. Welldon, 'Changes in the Population of Essex', *Essex Archaeol./Historical Society*, Vol. 4 (1972).

Fletcher, Sir Banister, *History of Architecture* (1956).

Forrester, H., *Medieval Gothic Mouldings* (Chichester, 1972).

Gethyn-Jones, Canon D., *Kempley* (1957). Available at the church.

Gibb, J. H., *Sherborne Abbey* (1972).

Hale, Archdeacon W. H., *The Domesday of St. Paul's* (Camden Society, 1858).

Harvey, Dr. J. H., 'The King's Chief Carpenters', *Arch. Jnl.*, 3rd series, Vol. XI (1948).

Harvey, Dr. J. H., *The Master Builders* (London, 1971).

Harvey, Dr. J. H., *The Perpendicular Style* (1978).

Harvey, Dr. J. H., *English Cathedrals* (1961).

Harvey, Dr. J. H., *The Medieval Architect* (1972).

Harwell, Berkshire Carbon 14/Tritium Dating Laboratory. Certificate dated 2 March 1976. HAR—1258. Age bp (yrs) 1020±90, bp — 1950, 930.

Hewett, C. A., 'The Barns at Cressing Temple, Essex', *Journal of Society of Architectural Historians*, March 1967, Vol. XXVI, No. 1. (U.S.A.).

Hewett, C. A., 'Aisled Timber Halls and Related Buildings, Chiefly in Essex', *Transactions*, Ancient Monuments Society, Vol. 21 (1976).

Hewett, C. A., and Smith, J. R., 'Faked Masonry of the Mid-13th Century at Nave-stock Church', *Essex Journal* (1972), pp. 82-85.

Hewett, C. A., 'The Carpentry', *Archaeology in the City of London*, No. 3 (1975).

Hewett, C. A., 'The Dating of French Timber Roofs', *Transactions*, Ancient Monuments Society, new series, Vol. 16 (1969).

Hewett, C. A. and Tatton-Brown, T., *Archaeologia Cantiana*, Vol. XCII.

Highfield, Dr. J. R. L., 'The Aula Custodis', *Postmaster*, Vol. IV, No. 4 (1970).

Hope-Taylor, B., (ed. by C. A. R. Rawley-Radford), 'The Saxon House, A Review and Some Parallels', *Med. Arch.*, No. 1. (1975).

Hunt, E. M., *The History of Ware* (Hertford, 1949).

Jervis, S., *The Woodwork of Winchester Cathedral* (Friends of Win. Cath., 1976).

Kent, Rev. J. A. P., *History of Selby Abbey* (London, 1968).

Kent, W., *The George Inn, Southwark* (1970).

Kidson, Dr. P., *A History of English Architecture* (1965).

Larking, L., *The Knights Hospitallers in England* (Camden Soc., 1857).

Leeds, E. T., *Early Anglo-Saxon Art and Archaeology* (Oxford, 1936).

Lingpen, A. R., ed., *Master Worsley's Book* (London, 1910).

McHardy, A. K., *The Church in London, 1375-1392* (London Record Society, 1977).

Mills, D., *Lambeth Palace* (Church Information Board, 1956).

Morant, P., *History and Antiquities of Essex* (London, 1768, Chelmsford, 1816).

Morgan, F. C., *A Short History of Abbey Dore* (1949-51).

Ogborn, M. E., *Staple Inn* (Institute of Actuaries, 1964).

Parsons, D., ed., *Tenth-Century Studies* (Chichester, 1975).

Pevsner, Sir N., *An Outline of European Architecture* (1943, 1963, 1973).

Pevsner, Sir N., *Buildings of England*—various dates of separate County publications.

Postan, M. M., *The Mediaeval Economy and Society* (1972).

Rackham, O., Blair, W. J. and Munby, J. T., *Med. Arch.*, Vol. XXII (1978).

Ray, Rev. P. W., *History of Greensted Church* (Ongar, 1869).

Repton, J. A., *Norwich Cathedral* (Farnborough, 1965).

Rickman, T. H., 'The Farmhouse, Thorpeacre, Loughborough', *Med. Arch.*, Vol. 12, (1968).

Rigold, S. E., 'Romanesque Bases, in and South-East of the Limestone Belt', *Antiquaries Journal*, Occasional Papers.

Rodwell, Dr. W., 'The Archaeological Investigation of Hadstock Church, Essex', *Antiquaries Journal*, Vol. LVI, part I (1976).

Royal Commission on Historic Monuments, Inventory, Essex, 1926. (H.M.S.O.).

Rudder, S., *A New History of Gloucestershire* (1779).

Salzman, L., *Building in England* (1952).

Sherley-Price, L., trans., *A History of the English Church and People* (1968).

Simpson, Sir J. W., *Some Account of the Old Hall of Lincoln's Inn* (Brighton, 1928).

Stenton, Sir F., *English Families and the Conquest* (1968).

Stranks, Ven. C. J., *Durham Cathedral* (London, 1971).

Taylor, Dr. H. M. and J., *Anglo-Saxon Architecture* (Cambridge, 1965).

Teledyne Isotopes, 50 Van Buren Avenue, Westwood, New Jersey, 07675. Dated 11 December 1978.

Isotopes Number 1-10, 488.

$-\delta\,C^{14}$	Age in years B.P.
112 ± 9	995 ± 80

Thorpe, B., *The Anglo-Saxon Chronicle* trans. Longman, Green (London, 1861).

Victoria County History of Essex (London, 1956).

Whitham, J. A., *The Church of St. Mary of Ottery* (1968).

Wilson, C. G., *Winchester Cathedral Record* (Friends of Win. Cath., 1976).

Wilson, D. M., ed., *The Architecture of Anglo-Saxon England* (1960).

Wood, R. G. E., County Advisory Officer, History; Essex County Council.

Introduction

A DECADE HAS ELAPSED since an introduction was written for my first book on structural carpentry, and since this is so short a time it is still appropriate to begin with the same remarks. There is a view, widely held, even among those having special knowledge of and interest in historic timber buildings, that carpenters had, at any time in the past, the whole range of timber joints known at the present time, from which they were free to select the one they preferred for their purpose. That is to say that the carpenter's craft, unlike any other, had had no history of development towards perfection; this is but one of a number of irrational opinions that gained a general acceptance in this country for reasons that are obscure. A similar view was that medieval buildings were not attributable to individual architects of genius, because 'Mediaeval men had an innate instinct for co-operative design' (Dr. J. H. Harvey, 1972, 9), or that man must have first built in stone. Such opinions probably originated in the higher centres of learning, since they obtained during a century of compulsory general education, but they are immediately ridiculous upon serious consideration.

The first to realise the differences of styles of carpentry, or styles of jointing and assembling frames, seems to have been Henri Deneux, of the French Historical Monuments Service. He published in July 1927 a volume entitled *L'Évolution des Charpentes du XIè au XVIIè Siècle.* This work should have revolutionised studies of historic timber buildings from that time onwards, since after reviewing over five hundred important structures he was able to show that : 'by examining all these examples of framework we have been able to prove, despite their great variety, that each period is characterised by definite assembly-methods (*dispositions*)'. It is now over half a century since this publication appeared, and its message is gradually coming to be recognised.

Deneux's methods cannot be improved upon, and the subject is approached by way of a selection of timber buildings that are representative of the technique of their times, reviewed at the end of each recognised period, and more synoptically reviewed at the end of the book. The periods of English architecture 'are based partly on historical periods and partly on architectural character', and 'as they have held the field for so long in all descriptions of English architecture, they have become, as it were, an integral part of architectural phraseology' (Sir Banister Fletcher, 1956, 347). In view of the inescapable truth of this statement these divisions have been used, with this overriding qualification—'What is important about the Styles is not their dates, which fluctuate (in different parts of the country) but their very recognisable general character' (J. H. Harvey, 1979, letter to author).

It becomes clear as the examples are considered that the effects of these changes, which were essentially stylistic, upon the works of structural carpenters are not

1

always obvious; and it also seems that certain styles directly affected structural thought, and not always to its mechanical advantage. The examples are described and illustrated, their structural joints and decorative features are listed in chronological sequence at the end of the book, and from these sequences the logical conclusion becomes clear—that carpenters' jointing techniques had of necessity to attain mechanical efficiency if their structures were to endure for any length of time. This is entirely obvious, and it was equally true of the masons' techniques, yet it has only been recognised in the second instance. It follows from this assessment of the historical succession of developments that timber buildings are datable by the techniques employed for their construction, a fact that has gained least acceptance among students of the subject, but which is gaining some recognition.

It is possible to establish when certain improvements were effected by studying numerous high quality buildings without regard to regional characteristics, thereby discovering what constituted the history of this particular technology at the national level—but it is difficult to show when out-of-date techniques lapsed into obsolescence. The use of jointing for the dating of buildings is therefore limited, but its value can hardly be over-estimated, since typological dating (which is our principal method) cannot be applied in part—that is, without regard to the most important type-series relevant to the study. Dating is fundamental to architectural history, and it is often problematical, since even incised dates such as occur on date stones or the side girt of Rooks Hall may record either the date of commencement or of completion. The same uncertainty attaches to cathedrals (which are generally the most fully documented buildings we have), because the records are primarily accounts, and as such concerned with the costs of the operations. The inception, possibly at Wells, of the single tenon mounted on spurred shoulders can only be broadly dated by Bishop Bubwith's will, but such records do assist with tolerably close approximations. It has in the past been considered that a building was dated if ascribed to a particular century, but the inadequacy of this is apparent when the present selection is considered—and the majority of works described cannot be dated with respect to each other. Two alternative and scientific methods of date determination are known, carbon[14] dating, and tree ring analysis— normally given the name of dendrochronology—and for a concise exposition of the principles of both the book by Professors Horn and Berger is recommended (R. Berger [ed.], 1970, 17-21 and 183-212). The first is a valuable but very expensive method, and the second is not yet available to the extent that is desirable. How useful these methods will become has yet to be established, and typological assessments that have due regard to the technological typology herein proposed are the best method available at the present time.

As in previous books it has been found most convenient first to describe and illustrate either the whole building in question, or such of its features as furnish evidence supporting the general thesis. This has been done with regard to the accepted periods of the successive styles of English architecture insofar as it has seemed possible, for not all timber buildings conform to those styles. At the end of each period the evidence provided by the selected structures is generally discussed, and after the selected buildings have all been introduced the evidence they provide is reviewed under the heading of a single category of joints or other building components, such as storey posts.

2

The important matter of cruck building has not been dealt with as those familiar with that building type would require; this is because only a few examples exist in the south-eastern counties. Some such buildings have, however, been mentioned as an aspect of the overall pattern. There has been only one historical course of development in English structural carpentry in the writer's opinion, and the preference for either straight or crooked timber was, irrespective of the regional pattern of distribution, one of the basic options that were open to carpenters. The use of base crucks cited in this text, in the capital and at relatively recent dates, endorses this view. It has been found in such cruck buildings as the writer has examined with the necessary thoroughness—such as the barn at Bishop's Cleeve in Gloucestershire—that the technological aspects of the carpentry, such as the scarfing, relate precisely to the general thesis expounded, and such buildings should be considered as an integral part of timber-building history rather than as something separate and distinct.

It has always been accepted that fieldwork can never be final, and while this text has been in production important new discoveries have been made which elaborate and confirm the thesis expounded; I have chosen to omit these, however, rather than alter the layouts and delay the production. In the event of a second edition many amendments will be appropriate, since this is a field of study that is in its infancy.

Chapter One

Examples from the Anglo-Saxon Period (A.D. 449 to 1066)

The Church of St. Andrew, Greensted-juxta-Ongar, Essex

This building is, in the present state of knowledge, unique in England. Its walls are made of oak trunks, split into halves and reared with their flat surfaces inward, a method of timber building for which the Venerable Bede seems to provide some evidence, referring to the year A.D. 664 (L. Sherley-Price, 1968, 185): 'He (Bishop Finan) built a church in the Isle of Lindisfarne suitable for an episcopal see, constructed, however, not of stone, but of hewn oak thatched with reeds after the Scots manner'. Furthermore: 'Eadbert, a later Bishop of Lindisfarne, removed the thatch and covered both roof and walls with sheets of lead', implying two important factors: a substantial building to carry such weight, and a metal technology able to produce sheet lead. Nothing else can be inferred from Bede, unfortunately, than that some churches were of timber during his time, i.e., the 7th century. Concerning the date of Greensted church there has been much speculation, and more doubt; but that the most acceptable of the half-logs in its walls are of a date between the 7th and the 11th centuries seems apparent, the terminal limit being provided by the translation of St. Edmund's remains in 1013. This event was recorded in a document of disputed date which is no later than *c.* 1300 (BM. Add. MS. 14847, f. 20. and *V.C.H.*, 1856, 60), and which has been much quoted: *Idem apud Aungre hospitabatur vero ejus nomine lignea capella constructa permanet usque hodie*—and seems to indicate Greensted church.

The association with Edmund, king and martyr, has clouded the matter of dating the structure, since it has been thought that it was built for the purpose of housing the remains, rather than that it was already in existence and at a convenient point along the route from London to Bedriceworth, afterwards Bury St. Edmunds. That it could not have been built hurriedly for the occasion is evident from its complex assembly method, and long-standing use of its site by the church has been established by a limited excavation, conducted in 1960 (H. R. Christie, O. Olsen and H. M. Taylor, 1979, 92-112). This was reported as follows: 'Recent excavations in the chancel have established the former existence of a small wooden chancel, of upright logs set in the ground, without any sill, and of a larger wooden chancel which replaced it; the larger chancel had a wooden sill and may therefore be assumed to be of the same general type as the existing nave'. This was approved, at the time, by all the excavators (Dr. H. M. and J. Taylor, 1965, 263). The last 'restoration' occurred in 1848-9, when it was thought necessary to rebuild the nave because the ground sills had rotted and the half-logs were said to be partially suspended from the top plates; a total rebuild was rapidly carried out, the *Builder* magazine of the day observing that this was of 'undue severity'. It certainly was, and the greater part of the western gable, which had survived until that date intact, was removed to provide an unnecessarily wide vestry door; whether the timbers were destroyed or went into some private collection has eluded the records of this disastrous event. A recent re-examination of what remains confirmed that all evidence for the original roofing method was lost at that time.

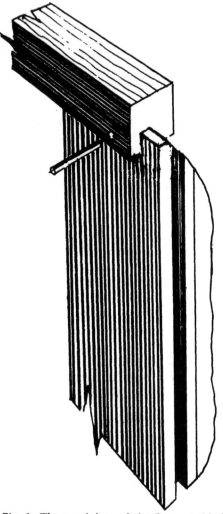

Fig. 1. The top joints of the Greensted half-logs with the top plate; perspective from Ray's drawing.

Study of the accounts that were published concerning this restoration serves only to prove disagreement between the various observers who wrote them, but these and the remains of the fabric are the only basis for a reappraisal of the building. The incumbent of the parish during the rebuilding was the Reverend P. W. Ray, who produced a small book about his church that ran to five editions (Rev. P. W. Ray, 1869, 19-20), in which he wrote: 'There are 24 timbers on the South side, and 25 on the North. The nave is 29ft. 9ins. long, 14ft. wide, and 5ft. 6ins. high to the top of the plate. The west end was carried up in the middle as high as the ridge of the roof, and consisted of two layers of planks fastened together with treenails. The planks are not long enough to reach the whole height, they are therefore so arranged as to break both the perpendicular and horizontal joints'. This information can be expanded by a more detailed account (Essex Record Office, Mint Portfolio, Greensted), which was also contemporary, and where accounts agree with each other *and* the remains of the fabric, grounds for deduction as to its original form and detail exist.

Another observer said of the walls: 'They are about 6 feet high, including the sill and the plate, and are formed of rough half-trees, averaging about 12 inches by 6 inches (the greatest length on the base line being 18 inches by 9 inches, and

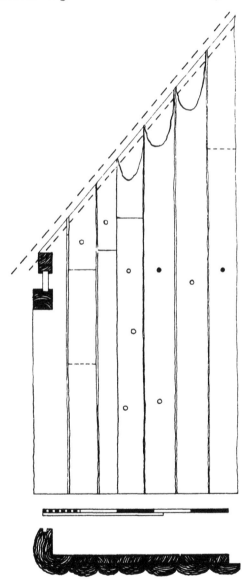

Fig. 2. Greensted, scale-drawn elevation of the north-west part of the gable, with horizontal section at base beneath the scale.

7

the least 8 inches by 6 inches). Mr. Suckling does not believe them to have been "half trees", but that "they had a portion of the centre, or heart, cut out, probably to furnish beams for the construction of the roof and cills; the outsides or slabs thus left being placed on the cills". We see no evidence of this, for the timbers were evidently left rough, and the dimensions prove them to have been, as nearly as may be, "half trees". These uprights were laid on an oak cill, 8 inches by 8 inches, and tenoned into a groove 1½ inches deep, and secured with oak pins. The cill on the south side was laid on the actual earth; that on the north side had, in two places, some rough flints, without any mortar driven under. The roof plates averaged 7 inches by 7 inches, and had a groove corresponding with the cill, into which the uprights were tenoned and pinned. The plates were also of oak, but they and the cills were very roughly hewn, in some parts being 10 inches by 10 inches, and in others 6 inches by 6 inches or 7 inches.

'There were twenty-five planks or uprights on the north side, and twenty-one on the south side. The uprights in the north side were the least decayed. Those on the south side required an average of 5 inches of rotten wood to be removed, those on the north about 1 inch only, and the heights of the uprights, as now refixed, measuring between plate and cill, are, on the north side, 4 feet 8 inches, on the south side 4 feet 4 inches, the cills being bedded on a few courses of brickwork in cement to keep them clear of damp. The uprights were tongued together at the junction with oak strips, and a most effectual means it proved of keeping out the wet, for although the interior was plastered, there was no evidence, in any part, of wet having driven in at the feather edge junction of the uprights,— a strange contrast to many of our modern churches, where, with all the adjuncts of stone and mortar, it is found no easy matter to keep out the driving weather from the south-west. The roof was heavy, and without any particular character; it consisted of a tie-beam, at less than 6 feet from the floor, with struts. The covering was tile'.

An alternative and conflicting account from the same portfolio (E.R.O., Mint Portfolio, Greensted) says: 'Each was cut with a tenon going into the cill; and the top of each was cut to a thin edge and pinned into the roof-plate'. This last detail was illustrated in several of the editions of the book by the Rev. Ray, and is redrawn as a perspective in Fig. 1. This is an important difference between the records quoted and the surviving fabric, which now has tenons along the tops; it is felt that these were probably cut from the thin edges by the carpenters affixing the new top plate. It is difficult to believe that the incumbent would have taken the trouble to have a block engraved and published in order to propagate a falsehood. There was, at the restoration, a south doorway, 4ft. 7ins. wide, with door-posts, and a northern doorway, 2ft. 5ins. wide—this last was found blocked and was reblocked with original timbers from some part of the structure.

The northern part of the west gable is drawn to scale in Fig. 2, in which a horizontal section showing the assembly testified by Ray is drawn at the base. No part of this wall that can be examined today confirms Ray's statement, but outside measurements set against inside measurements taken from the actual corner suggest the construction drawn—which, however, cannot be proved. In fact, at least two half-logs with grooved sides can be seen from within the west tower, one of which is scarfed simply and was fixed with two nails. This area, when checked and compared with the complementary south-western area, shows enough original peg holes to form a horizontal line across the wall at a height that was probably between the two top plates at each side; this indicates the original existence of a tie beam at this end, without which the whole would have been too flimsy to survive.

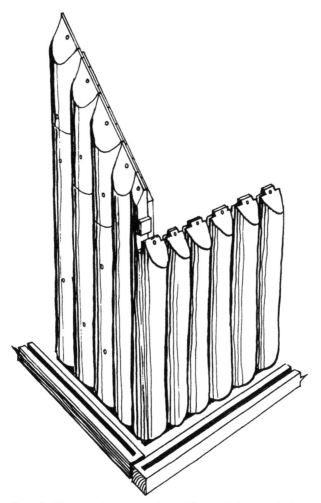

Fig. 3. The south-west angle at Greensted, raised above the grooved sills.

The south-western corner is drawn as Fig. 3, in which the 'thin edges' shown by Ray and other contemporaries are drawn. Both western corner posts have a quarter removed to form their internal angles, and this is a feature which no contemporary plans show, but on-site examination has failed to prove that the northern post is not an original, due allowances having been made for the partial replacement of its timber at the top. The most curious detail of these corners is the elaborate arrangement for mounting two top plates, each rebated to carry a plank-on-edge, between them; it is difficult to envisage a Victorian carpenter inventing this elaborate system, and one suspects that it is a reflection of the original. The original uprights of this end are taller than the corner posts by the corresponding amount, a fact supporting the authenticity of the method.

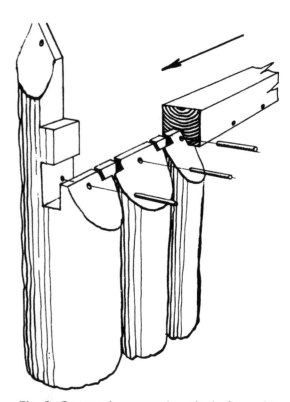

Fig. 4. Greensted, method of assemby against the corner logs.

Fig. 5. Greensted, suggested method of top plate insertion.

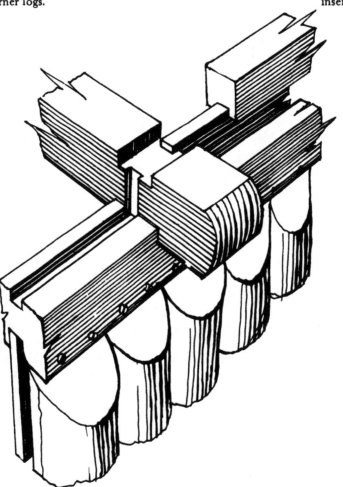

Fig. 6. Greensted, hypothetical reconstruction of the missing central tie beam.

The timbers shown in Fig. 2 with their tops cut obliquely flat have survived in that form, and suggest that the verge rafters were set on to the same form of tenons as were used for the side walls. As shown in Fig. 4 the studs (half-logs) probably had to be reared against the corner posts, into which the fillets would have been previously inserted with respect to the side walls, the same corner logs being rebated to accommodate the end wall. The top plates could then have been advanced, endwise, into the rebates cut in the corner posts (Fig. 5). Alternatively, all the side walls complete with sills and top plates, pegged together and each with terminal corner posts, could have been reared by sheer manpower, and the whole stabilised by the tie beams at that point. The method of fitting the tie beam shown in Fig. 6 was, in fact, hypothetical—but the most recent on-site examination proves that this was the method used by the 1849 restorers. It is probable that a tie beam

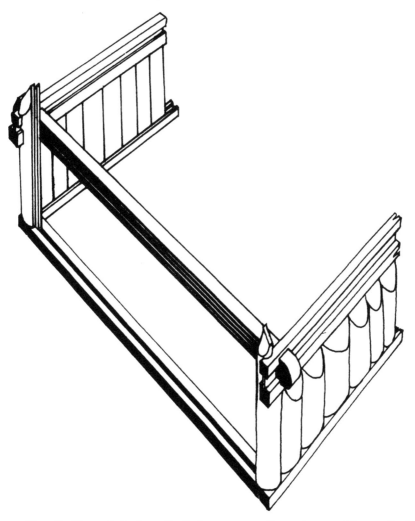

Fig. 7. Greensted, hypothetical reconstruction of assembly method
for the western end, against the tie beam.

11

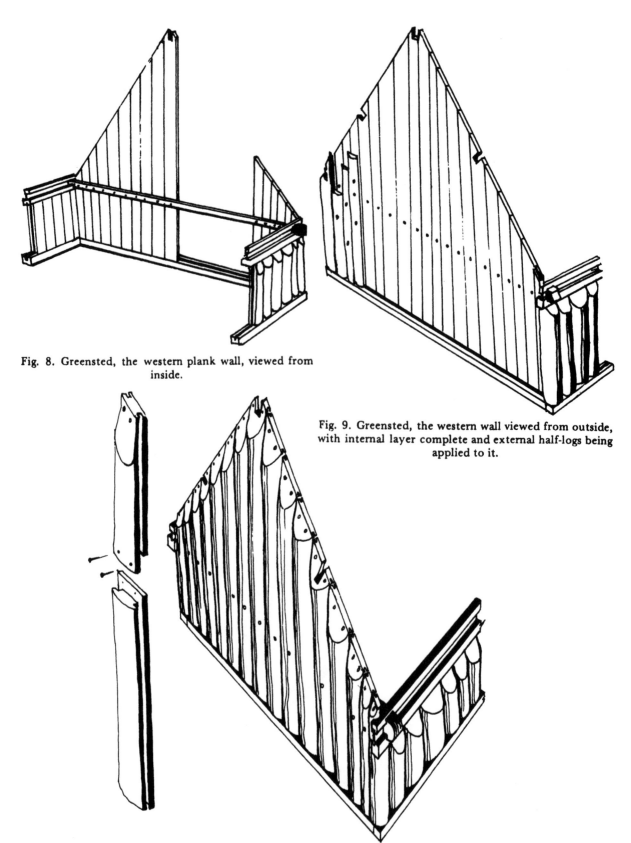

Fig. 8. Greensted, the western plank wall, viewed from inside.

Fig. 9. Greensted, the western wall viewed from outside, with internal layer complete and external half-logs being applied to it.

Fig. 10. Greensted, reconstruction of the complete western gable, with exploded view of joint in a half-log.

was fitted at the eastern end, but nothing survives there today. The former existence of the western tie beam is not only substantiated by the peg holes now approximating to its line; it was also obviously necessary because if, as Ray described it, it was built of two layers all vertically placed and slightly overlapping, it could have possessed absolutely no strength with which to survive until 1849. Fig. 7 shows such a tie beam. Against this beam the western wall could have been reared, either of the two alleged layers or of single half-logs—a structural point upon which the visible evidence is equivocal. Figs. 8, 9 and 10 illustrate possible processes for this operation. Fig. 10 shows the west as though built of grooved half-logs, together with the scarf visible on one half-log that survives in the tower at south-west enlarged on the left. Fig. 11 is a purely hypothetical impression of the whole, with assumed ridge piece and purlin slots.

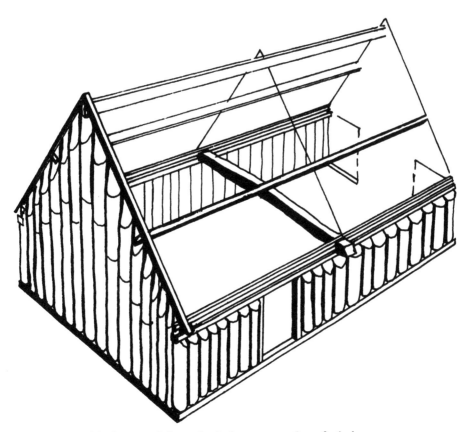

Fig. 11. Greensted, hypothetical reconstruction of whole carcase.

The plank-on-edge system of wall plating perpetuated by the 19th-century restorers is accounted for in Fig. 6, wherein it may be seen that this plank fills the gap betwixt the top plates which was caused by fitting tie beams between them. These suggestions go a little way towards understanding what little remains at Greensted, and indicate a remarkably close relationship with the grave chamber of the Gokstad ship.

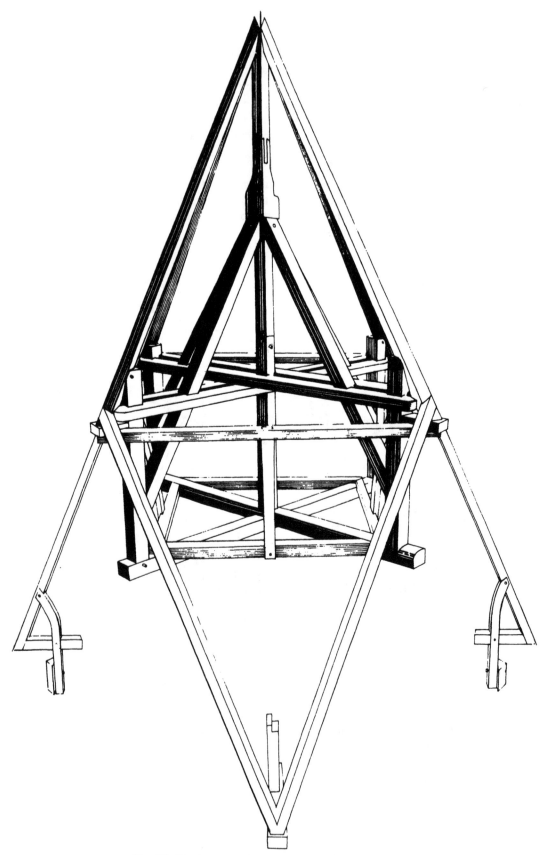

Fig. 12. The framing of the Rhenish helm at Sompting.

The western tower of this church retains a rare example of the spire form known as the 'Rhenish helm', a square-planned and pyramidal spire that rises from the apexes of four gables. This is a survival that has generated many 'scholarly' mis-interpretations, including a tradition that its height was reduced in 1762, and another, more recent and even less probable, that it was rebuilt after the Conquest. The present state of knowledge does not admit of a date ascription closer than somewhere between *c.* A.D. 950 and *c.* 1050. It is unlikely to be later than this.

No other Anglo-Saxon tower in this country has a contemporary spire of this type, but there are structural indications that St. Benet's, Cambridge, was originally of the same form; and it is suspected that others were replaced by more conventionally English types right up to the 19th century. The architectural and structural concept of the Rhenish helm is extraordinary, but its execution in carpentry at Sompting is a work of such assurance and competence, achieved with such economy of means, that it both indicates the work of a master and suggests the previous existence of a tradition of framing such works.

The entire framing of the helm is shown in Fig. 12, in which enough components have been omitted to clarify the matter. Until such time as the cladding is removed for repairs there are points of construction that cannot be verified, and these are omitted. They include the framing of the purlins to the eaves rafters, and the 'valleys'. It is apparent from both structure and situation that the helm had been previously worked and fitted together at ground level, when it was numbered for reassembly with chisel-cut Roman numerals that are still clearly visible. Late Saxon method, therefore, anticipated ensuing carpenters' methods. The verges of the gables were not finished in the level plane, but were inclined at the same pitch as the surfaces of the spire, for which reason the gables' truncated tops produced triangular flats.

The whole structure was soundly designed, as its survival proves; and the free-stone columns formed by the quoins and the mid-wall pilasters were placed under the heaviest loads, the helm being designed to stand on eight bearings, four at the gable apexes and four at the valleys, and to have a central steady in the crossing of the transoms, in which the spiremast was seated. This system produced a spire that wind pressures have failed to dislodge from its tower. The first unit of framing assembled in the tower top is shown as Fig. 13. The diagonal braces fitted in the horizontal plane would have been fitted next and jointed (as shown in Fig. 14) with chase tenons and, more important, modified lap dovetails in pairs and addressed. This assembly greatly stabilised the four gables and provided locations for the feet of the four principal-rafters and the central spire mast: the latter was lowered through the two square voids formed by the four rising braces and the crossings of the paired straining timbers, when its foot tenon could enter the mortise on the upper transom (Fig. 13).

Structural details of the four gable posts are given as Fig. 15, in which the direct and simple techniques are of interest; for example the arris trenches cut into the posts' tops to receive the principal-rafters are unique to date in the writer's

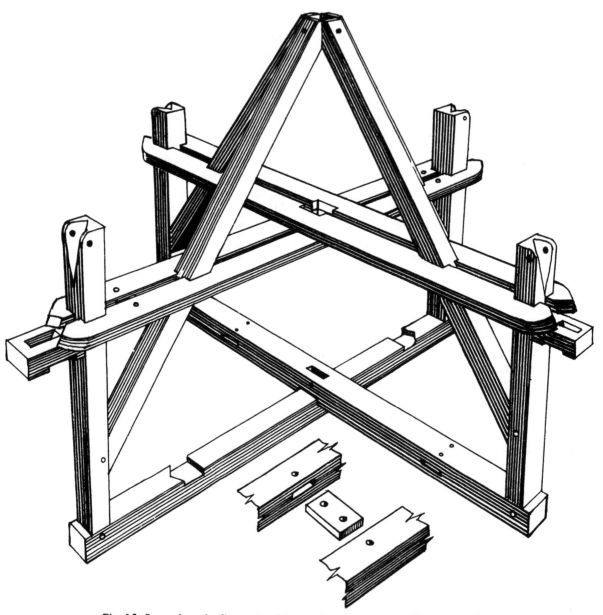

Fig. 13. Sompting, the first unit of the timber framing, with 'free' tenon inset.

experience. When assembly had reached this point the spiremast would have been fitted (the two unrelated kinds of scarfing it shows will be discussed elsewhere), and when positioned it was face-pegged (Fig. 16). This last is a recognisably Saxon technique, and one which can be traced through the 'Saxo-Norman Overlap' and into the 13th century. Note the hewn outsets, on the faces of the mast, designed to rest on the tops of the braces. Hereafter the four rafters forming the arrises of

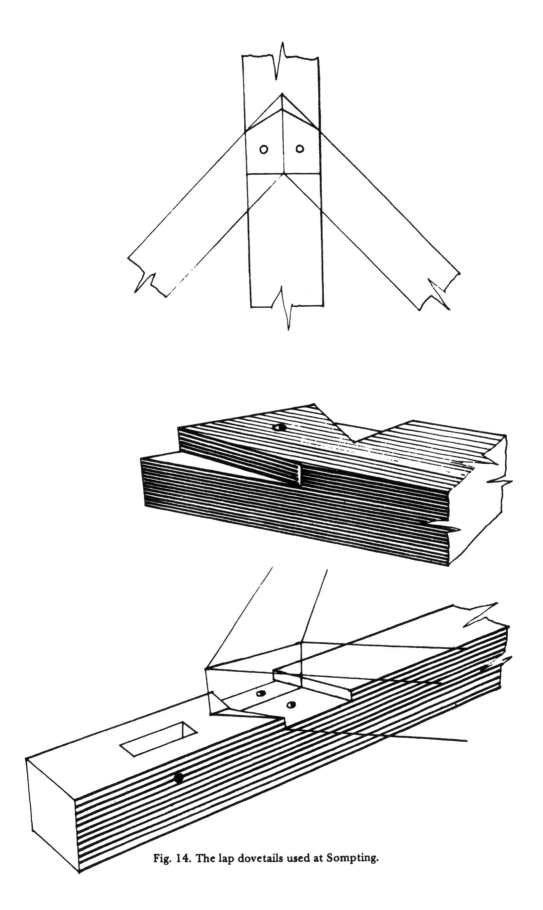

Fig. 14. The lap dovetails used at Sompting.

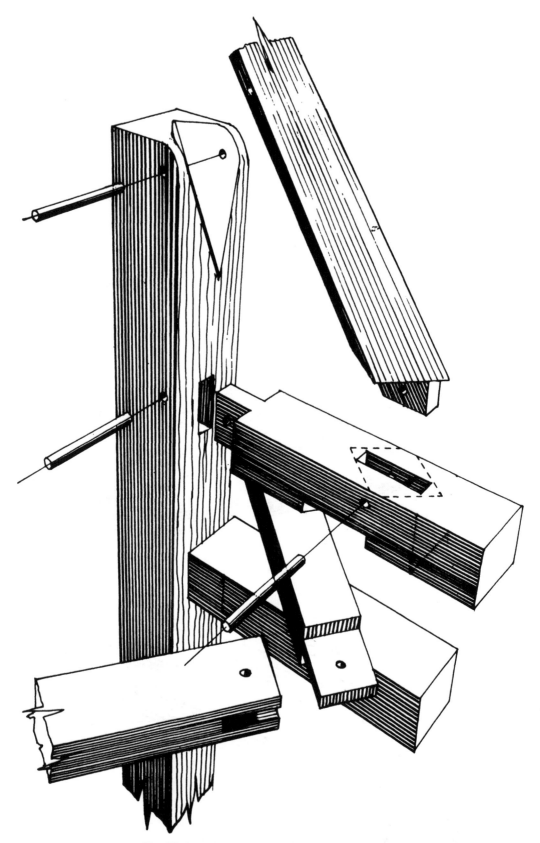

Fig. 15. Sompting, exploded view of one gable post.

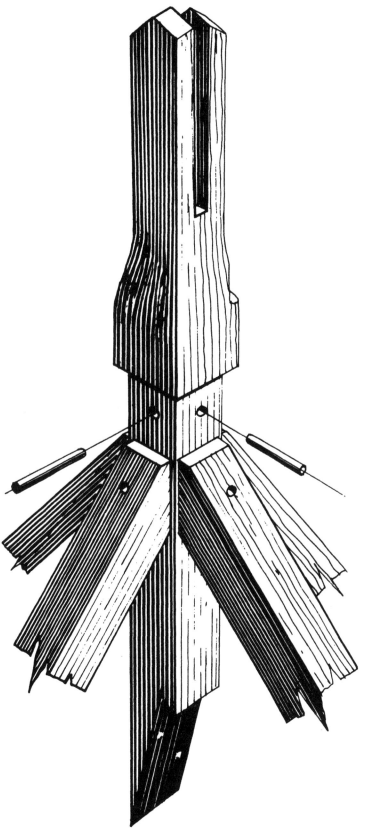

Fig. 16. Sompting, central spiremast section, showing both scarf joints
and outsets with face-pegging.

the helm were fitted, and then the common rafters, five to each facet, each being made of two lengths. The central rafters which run diagonally down to the lowest points of the helm were wall-anchored (Fig. 17). These are unusual examples, in timber, of functional members that were later made in wrought iron. They presage the base triangulation of common rafters which was to become invariable during the medieval period; and the lap joints used to make them, whilst adequate for their purpose, were unable to resist withdrawal and may be the origin of the later notched lap joints.

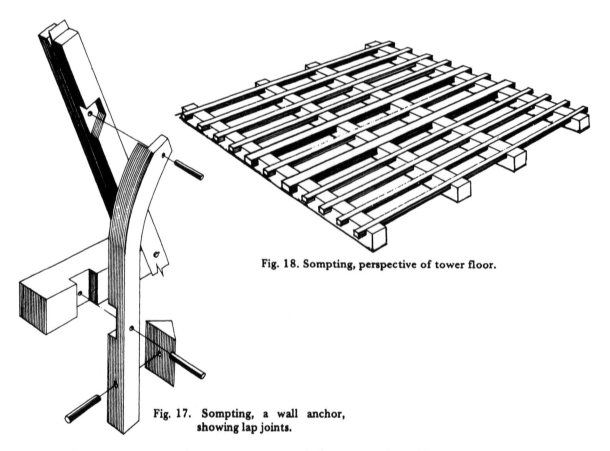

Fig. 18. Sompting, perspective of tower floor.

Fig. 17. Sompting, a wall anchor, showing lap joints.

The framing of the floor beneath the bell chamber is also of interest (Fig. 18), and appears to be original to the tower. Unlike other floors this was laid upon four bridging-joists aligned north to south, and spaced in such manner as to leave a central interstice equal to the diameter of the largest intended bell. Across these, and apparently trenched into their upper faces, were laid two more joists of a cross section intermediate between the bridging-joists and the common joists. The latter, insofar as has yet been determined, were located in cross coggings.

20

The Church of the Holy Trinity, Colchester, Essex

The western tower of this church has many features indicating a date shortly before the Conquest (G. Baldwin Brown, 1925, 447; H. M. and J. Taylor, 1965, 143-5). The tower and its internal timber work make a confusing spectacle from within, giving the impression that the 18th-century timbers were inserted to support the floors; but closer examination proves that they were added to provide vertical stability to the rubble walls, which have dressings of brick, some of which must be Roman. The points at which the joists enter the walls seem mainly undisturbed, and if this is accepted then the floors must be considered original and late Saxon. Their framing does not resemble the Sompting example and has one heavy bridging-joist from east to west, across which are laid eight common joists. All timbers are squared to Roman imperial measurements, the biggest being 12ins. square, with a two-inch stop chamfer along its soffit arrises, whilst the common joists measure 7½ins. deep by 8ins. wide—a section presaging the later medieval use of timber laid 'flat', instead of on-edge. This floor merits no illustration, being simple, as described.

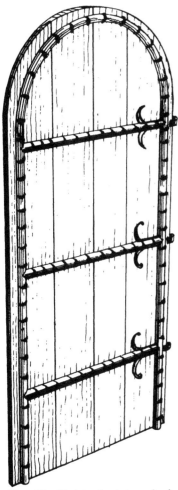

Fig. 19. Hadstock, internal view of north door.

The Church of St. Botolph, Hadstock, Essex

Recent excavations carried out by Dr. W. Rodwell have established that this church has passed through at least three phases of building, the last of which was during the 11th century. The fabric retains at least five pieces of Saxon carpentry—four mid-wall window frames and its famous north door leaf. The latter has long been accepted as Anglo-Saxon because it is Romanesque in style, but has no structural affinities with Norman door-leaves, a larger number of which have been identified. This is shown as Fig. 19. It was constructed from four wide planks which must have been seasoned at the time of use, because the long edge joints spaced by ironwork have never opened due to shrinkage; the joints are splayed rebates. The ledges forming the rear frame are made of D-sectioned oak, the top one being bent to the arcature of the door's head. The whole is fastened together by means of iron roves and clenches, the roves being so elongated as to encircle the wood and prevent it splitting when the clenches were formed (Fig. 20). These techniques all derive from ship-wrighting, and have survived in use until today, when they may be seen in the building of clincher-built small craft.

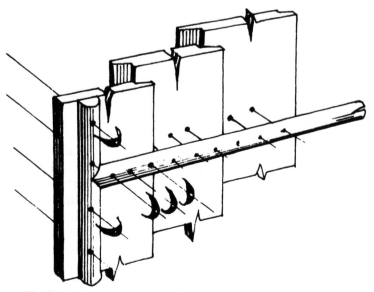

Fig. 20. Exploded view of the Hadstock door assembly.

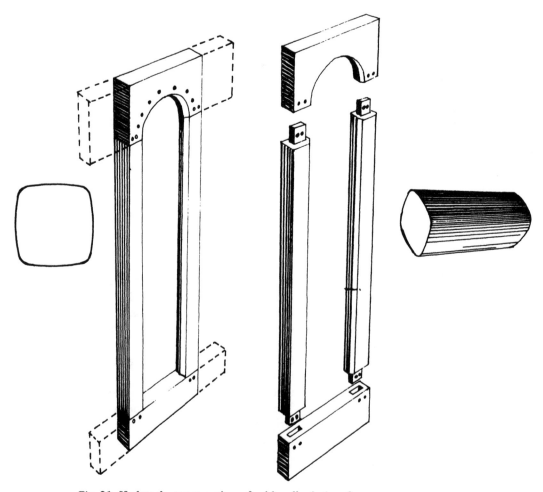

Fig. 21. Hadstock, construction of mid-wall window frames.

The mid-wall window frames, four of which survive to be seen, are high up in the double splayed openings of the nave walls; they are the only recorded examples of the kind that were constructed from four carpentered timbers, jointed together and pegged (H. M. and J. Taylor, 1965, 676-7). One of these is drawn as Fig. 21. They exhibit carpentry that was skilled, and each have two chamfered stiles, a sill and an arched head timber, assembled by stub tenons transfixed with two pegs each.

The head-and-sill timbers penetrate the rubble walls for a considerable distance, as indicated by dotted lines in Fig. 21. I am indebted to Dr. W. Rodwell for this information. The pegs are of the curious section illustrated at left and right of the drawing. The cutting of such holes suggests the use of three tools: an auger, a scribing gouge of the sides' curvature, and a second scribing gouge to cut the rounded 'corners'. A close study of this work proves that these tools were of good steel, very sharp, and handled by a craftsman proficient in their use.

The barn at Paul's Hall, Belchamp St. Paul, Essex

This is one of the ancient manors of the Chapter of St. Paul's in London, regarding which remarkably detailed farm leases have been preserved. Some of the leases— no more than a sample—were published by Archdeacon Hale, in the case of this holding a lease from the time of Dean Hugh de Marney, 1160–81 (W. H. Hale, 1858, 138-9). The surviving barn is the last of three that existed within living memory; it is aligned from north to south and sited west of the house, whereas the lease describes only two barns, both of which were aligned east to west. The given measurements in the lease, however, are not irreconcilable with this barn, which may therefore have been moved when rebuilt in *c.* 1200, as there is structural evidence to show. It is this survivor that incorporates the features most important for the historian of timber building.

The whole has been rebuilt more than once and greatly extended in length, some of the later works including re-used timbers, as a result of which it contains a diverse series of 'developmental' features. At its northern end are what appear to be the remains of a three-bay building that used structural principles more archaic than any hitherto recorded. The only undamaged post at this end stands closest to the south-west angle of the barn, and is earth-based upon a lime/cement pad placed 6ins. beneath the floor. Its foot is completely independent of the side wall behind it. This post is scale-drawn as Fig. 22, wherein absent components are drawn in chained line. It has been established by a small excavation that the post was moved at an early date and re-erected in its present position. It seems from a packed layer of pebbles that at that time the remnant of the earthfast shore, its foot rotted off, had been left *in situ*. A carbon[14] date was obtained for this post (for which I am indebted to B.B.C. television's *Chronicle* programme) and the determined age was 924 years, plus or minus 95 years, before A.D. 1950. The central date is therefore A.D. 1026, and the dating range is 931 to 1121. It can be proved that the existing tie beam is later than the post because the empty mortises in the post have no equivalents in the beam, which together with the existing top plate point to major rebuilding, *c.* 1200.

This example provides a pair of late Anglo-Saxon posts belonging to an aisled timber building (which was probably a barn since it was never sooted), one of

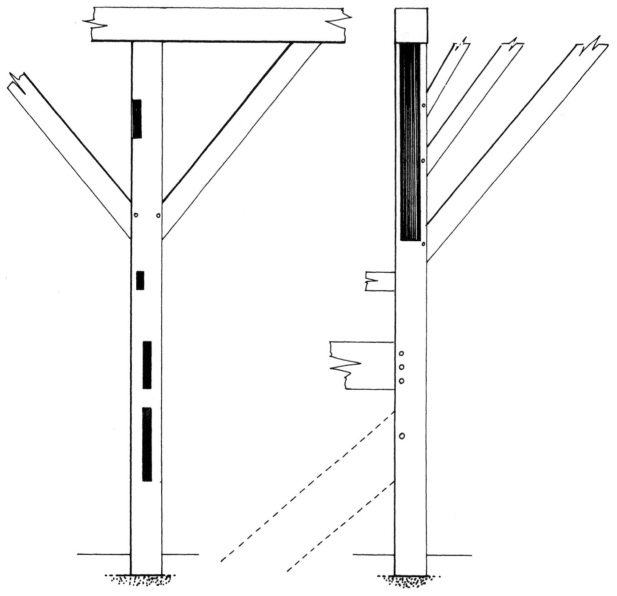

Fig. 22. The barn at Paul's Hall, scale-drawn rear and side elevations of a post.

which retains a basal standing of a probably pre-Conquest type. It suggests the existence of elaborate aisled buildings which possessed many of the structural features, such as passing braces, that were to predominate for two centuries after the Conquest. Furthermore, such structures if razed from their sites would have left evidence neither of their elaborate designs nor of the skills of their carpenters.

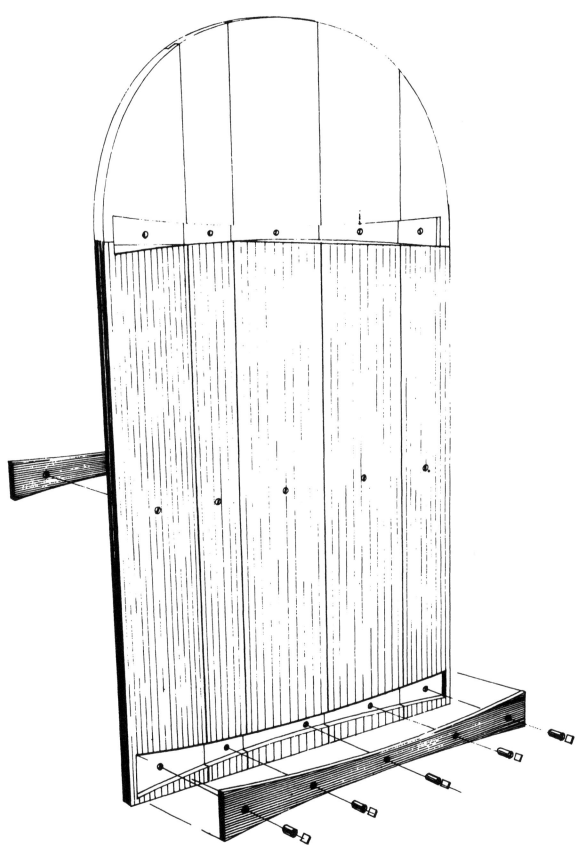

Fig. 23. The Westminster Abbey Saxon door leaf.

Door leaf in Westminster Abbey, London

This is illustrated in Fig. 23 as partly reconstructed, the hatched area representing all that survives intact (which is the larger part of the artefact). It has been suspected in the past that this was once the door to the Pyx Chapel of the Confessor's building, of between 1055 and 1066; it is now relegated to a position behind the bookstall in the eastern range of the cloister. The construction is remarkable, and no comparable door construction is known to exist elsewhere at this time. It is formed of five oaken planks with squarely rebated edges, strengthened in-plane by three ledges which were let into housings worked on opposite faces of the planks. Each ledge was cut into a doubly dovetailed shape, the lines of their divergence being arcs of great circles; the final fixing was by pegs, the ends of which were wedged (Fig. 24). The accuracy of the housings cut into the plank faces indicates the use of good, sharp-edged tools, which must have included a router and some type of cutting gauge to cut the radii of the sockets' edges to depth.

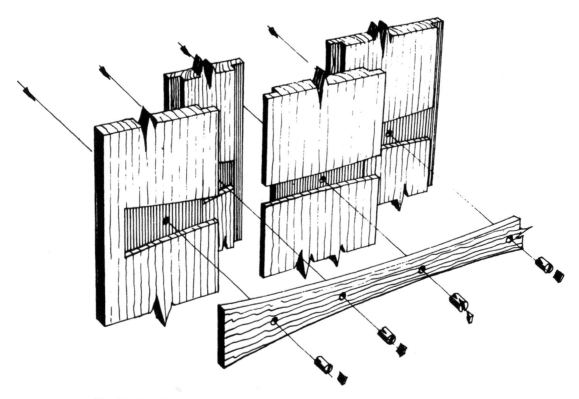

Fig. 24. The Westminster Abbey Saxon door leaf; exploded view of its assembly method.

The Church of St. Mary, West Bergholt, Essex

This church has recently been renovated, and an excavation concurrently implemented by the County Archaeological Section indicated that during the 14th century major rebuilding had occurred to an earlier fabric, which had itself been preceded by a single-cell Saxon church with an apsidal eastern end. The north wall of the Saxon building survives in the existing north wall; the excavator assessed its date as *c.* A.D. 1000, and was interested by a complete lack of evidence for a west wall to the church. This suggested the former existence of a western timber structure, which on removal had left no trace of its existence.

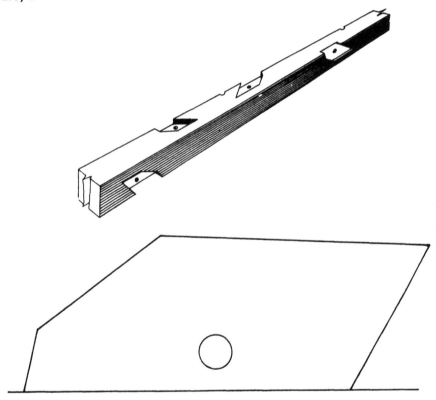

Fig. 25. One of the re-used timbers in the belfry at West Bergholt, with one lap joint matrix enlarged.

Several ancient and obviously re-used timbers were found in the existing belfry, examination of which suggested that they might be salvaged from the demolished timber west end of the church. These timbers are of oak, short-grained and heavy of section in relation to their length, and were evidently a low-cost makeshift to provide a floor in the timber tower which dates from the 14th century, and possibly from the same rebuild as has been previously mentioned. One of these is illustrated as Fig. 25, and a photograph of another forms Plate I. Several of them show matrices for squinted notched lap joints that occur on adjacent faces of the same timbers, suggesting that they were the corner posts of a tower structure similar to that at Navestock, Essex, or Pembridge, Herefordshire.

The Anglo-Saxon Period—Summary

From the eight works described it is possible to adduce a number of facts concerning carpented buildings during the period, with, of course, a heavy emphasis on the close of the Late Saxon period. This information probably covers, thinly, the 9th, 10th and 11th centuries. It is possibly adequate in substance to dispel the curious view, published only four decades ago, that the 'bulk of the people, we can now be assured, were content with something that hardly deserves a better title than a hovel, only varying in its greater or lesser simplicity' (E. T. Leeds, 1936, 21).

Archaeologists have in general regarded this view with suspicion, since the evidence of excavated sites could seldom be reconciled with its conclusion; but no works in timber could be shown to pre-date the Conquest, and the traces left in site surfaces by earthfast posts do not assist with meaningful reconstructions. Neither, had any accuracy been the unlikely product of the exercise, could any grounds with which to defend it have been produced. The basic concept of earthfast posts today seems archaic, even primitive and ill-informed; but this view is difficult to support with reasoned defence. To drive a post into the site of a building, or to embed it therein in any alternative way is the most reasonable way of fixing it that comes to mind, and the present-day illustration of this fact is provided by the field gate post, properly named a bar post. These are now, as always, earthfast; because no alternative or better method of placing them is known. Fixed in this way they have an average life, it is generally affirmed, of half a century. Such posts can leave no subterranean evidence concerning their form above ground, nor anything to indicate that a five-barred gate was suspended from them. Researches in this sphere have attempted to deduce the known from the unknown, which has proved impossible. The alternative method, enunciated as long ago as Rickman, for the isolation of building types demonstrably preceding other known types, has proved capable of determining which carpentry pre-dates the Conquest by reversing the accepted process, and differentiating the unknown from the known.

The literary references from Bede and Beowulf are of interest and appear, in the light of excavations, to be reliable for a great deal of their architectural detail. Bede, for example, states of Aidan (A.D. 651): 'As he drew his last breath, he was leaning against a post that buttressed the wall on the outside. He passed away on the last day of August, in the seventeenth year of his episcopate . . .'. The church referred to was twice destroyed by fire after this event, when it was alleged that due to a miracle the post on which Aidan had leant was preserved from the flames; 'for although in a most extraordinary way the flames licked through the very holes of the pins that secured it to the building, they were not permitted to destroy the beam'. Although the translator cannot be excused for confusing a beam with a post, or a buttress with either, this is contemporary evidence for the use of such buttresses as have been excavated (C. A. R. Radford, 1957, 17–38). The description in Beowulf of the wine hall is considered credible, but it seems strange that the 'iron clamps, forged with curious art', that so adequately strengthened it have not been excavated as yet. No reason is known for the 'interpretation' of these accounts, other than the linguistic necessity; and it is considered that wine halls were probably both 'splendid and horn-bedecked', containing 'many a mead-bench adorned with gold'. That the patterns of post holes on excavation sites cannot provide this much proof is not good cause to doubt the contemporary evidence.

Concerning the church at Greensted a mid-9th-century date was determined (this is published in the current guidebook to the church), allegedly from the annual rings of the half-logs where they are splayed at their tops. Irrespective of this date or the method of its determination, a date relatively close to that of Bishop Aidan's episcopal church in Lindisfarne is credible, since Bede makes it clear that such buildings were temporary, and were replaced or otherwise improved as soon as was convenient. From the limited excavation undertaken in 1960 (H. R. Cristie, O. Olsen and H. M. Taylor, 1979, 92–112), little evidence was forthcoming, save the statement already quoted; and the traces of earthfast posts from the earlier chancel cannot be interpreted as being either for circular or semi-circular earthfasts. It is possible, furthermore, to interpret Bede's *de robore secto totam composuit* as specifying half-trunks, or merely 'cut' trunks; although the latter interpretation renders the statement superfluous since you cannot build with trunks that are not cut: and it is suspected that cutting into halves lengthwise may be intended.

As is shown in the hypothetical reconstruction, this building has some merit, and the 19th-century comments as to the extreme effectiveness of the filleted long joints, betwixt the uprights, suggest the skill involved in the execution of that part of the work. It is considered unlikely that those uprights (whatever their cross sections may have been) would have been re-used at such a time as groundsills were fitted, for no craftsman ever liked using old material for a second time, no matter how frequently lack of funds may have occasioned it. The most important fact to be learned from the Greensted church is that oak, if so allowed to do so, will survive outdoors for a thousand years, and survive in a condition to defy saws, as the 19th-century restorers recorded. There is, as a result, no limit as yet determined for its survival inside a building.

A structure bearing striking resemblances to Greensted church is the burial chamber excavated with the Gokstad ship, which was also built of vertical logs with their inner faces flattened to varying degrees. Being basically a roof structure that was designed to lift and fit on to the ship, it was base-tied, but the ties in that context amount to sills, being the basis of the structure. The Gokstad Ship was built·in *c.* A.D. 800 (*Vikingskipene*, 1970, 18), and was interred within a century, which establishes a relationship betwixt the light, strong and elegant ships and the massive log-building tradition of Greensted. The Gokstad grave chamber is the contemporary land building, without modification, and was not a deckhouse in the maritime sense. That these two structural methods should be contemporary is reasonable because there was no obvious limitation on the dead weight of land buildings, while that of ships was determined by the buoyancy that was essential; and it would seem that each of these traditions could have influenced the other, as the rebated clincher-build of the Hadstock door establishes.

The types of carpentry at Greensted and Sompting are so different as to suggest a long lapse of time intervening between the two; the conjectural dates for each— *c.* 850 and *c.* 950–1000—only admit to two centuries and imply an earlier date for the log-church. The structural method at Sompting is competent, and the workmanship wrought with an assurance that must indicate the previous existence of a long tradition. The floor described with regard to Sompting incorporates re-used timbers showing joints not otherwise present in the tower, two of which are

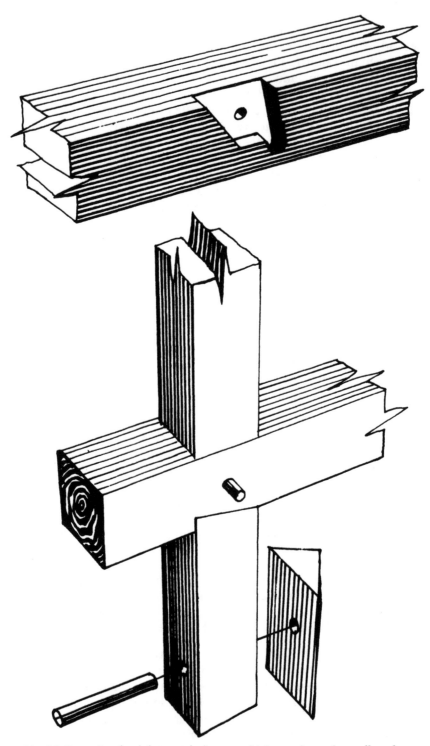

Fig. 26. Sompting, lap joint matrix in re-used joist, and wooden wall anchor.

illustrated in Fig. 26. The upper example is a squinted lap joint that was carefully cut into a 'nosing' such as was normally used to resist compression, and the date of this is uncertain. The lower drawing shows the wall anchor in the south-western valley of the spire, a cross halving with at least one spurred shoulder. Another technique that has hitherto been considered to originate during the 13th century is shown in Fig. 27, a tenon with one shoulder 'scribed' to fit over a waney edge. The sum of techniques used at Sompting provides a basis for the development of the majority of carpenters' joints, in the light of which it is no longer possible to ascribe the introduction of any types to the Conquest, or the Normans.

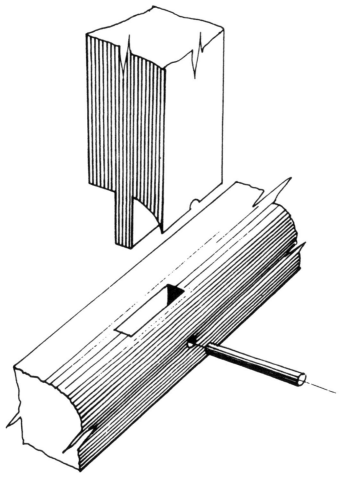

Fig. 27. Sompting, scribed shoulder to central tenon.

The uniqueness of the Sompting tower floor is probably significant for other floors of the period, and large examples forming western galleries which were able to support large numbers of congregations have been shown to have existed in Saxon churches; one such was at Deerhurst (H. M. Taylor, 1975, 164), but none have left structural evidence as to their framing. Concerning such floors, however, the *Anglo-Saxon Chronicle* records one that collapsed in the year 978: 'In this year all the Chief 'witans' of the Angle race fell at Calne from an upper floor ('de solario converunt'), except the holy archbishop Dunstan, who alone was stayed upon a beam; and some there were sorely maimed, and some did not escape with life' (B. Thorpe, 1861, 99). This disaster caused national concern, and the floor, which must have been large in area, would have been redesigned, and refinements that eventuated would have passed into wider usage among carpenters as a result of these aristocratic injuries and deaths.

The Sompting floor may pre-date the recorded disaster, and the next described, at Holy Trinity in Colchester, is for numerous reasons ascribed to a later date. Both are merely tower floors of small plan areas, but it is possible that in their structural differences they reflect a widespread advance in flooring, prompted by the historic disaster. The Colchester floor, which is the lower of the two there, employs joists lodged upon the bridging-joist and has the ends of all timbers embedded in the masonry. This is so direct as to be foolproof, and all components must function fully in load bearing; the Sompting specimen was more ambitious and sought to produce a flat surface above and beneath at the expense of the timbers' compression stria. The latter also evinces a concern on the part of the architect for the appearance of the underside of his work, which is elegant and unusual if compared with that of Holy Trinity; and a possibility of foreign influence may be suspected so near the Sussex coast via trade with the Rhineland. A recent book has pointed out that Anglo-Saxon trade with the lands of the Rhine estuaries was extensive throughout the period (M. M. Postan, 1972, 208–211).

Hadstock church has much-debated associations with both St. Botolph and King Cnut, the former having founded a monastery that has yet to be located, and the latter having founded 'a minster of stone and lime' in the year 1020; there are grounds for assuming that this church may be the minster, and that it may also contain the site of the saint's first burial (Dr. W. Rodwell, 1976, 55–71). Apart from its associations this remains an undisputedly Anglo-Saxon church of considerable importance, and formerly elaborate plan. The skilled carpenters' work exhibits the differences of principle that are necessary to establish it as a Romanesque precursor of Norman carpentry, while its advanced and lightweight use of material make an early 11th-century dating acceptable.

The door leaf in Westminster Abbey illustrates a further advance in the delicately-carpentered style that all these late examples illustrate, a style that probably did not lend itself to any great increases of size.

The barn at Belchamp St. Paul's is dated as being nearly contemporary with the last Saxon phase of the church at Hadstock, and among the buildings described it affords more evidence than any of the others concerning the least-known field of pre-Conquest studies—that of aisled timber-framed halls. For this reason as much reconstruction as its surviving timbers can support is justifiable. It is probable, in the light of a limited excavation, that the building was moved at a time when the earthfast part of the post shores had rotted off, a fact that also accounts for its different orientation from that of the lease, which it pre-dates. What little survives gives us an informative and immediately pre-Conquest example of an aisled timber building that relates closely to most post-Conquest examples of the same type.

The evidence includes firm indications that several features, such as passing braces and reversed-assemblies, have origins centuries earlier than has been supposed. The most important detail of these posts is the chase mortise near the base of the rear face of each, by which the posts could be reared against their earthfast shores, and then secured with pegs. It is apparent that these shores were of the same section as the posts, and it is suggested that they were set into the site by means of ramped post holes, neatly cut, and subsequently back-filled (Fig. 28).

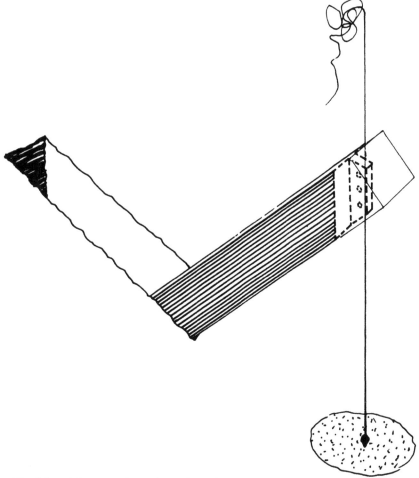

Fig. 28. Belchamp, a hypothetical earthfast shore in a ramped pit, with plumb line to cut chase tenon at top.

On a clay site this method would have produced an immovable buttress, at the cost of minimal labour with a spade; the chase tenons would then have been cut, accurately positioned by means of the plumb bob illustrated. A hypothetical site with four posts reared in this manner is drawn as Fig. 29. It is possible that this technique derived from the system recorded by Bede with regard to the death of Aidan, in which external timber buttresses reached to eaves height; examples of this have been proved by excavation in the case of 'The Hall of King Edwin' (B. Hope-Taylor, 1976, 69). As was true of all timber-frame rearing processes a well-rehearsed and efficient procedure had to be employed in order that stability could follow immediately upon   rearing, since that stability always results from the collective sum of the components. Once set up against their shores the posts must have been promptly incorporated into the evidently complex building they supported. It is appropriate, at this point, to reiterate that most basic types of carpenters' joints had been used before the Conquest, and that a variety of sharp tools was available to the craftsmen.

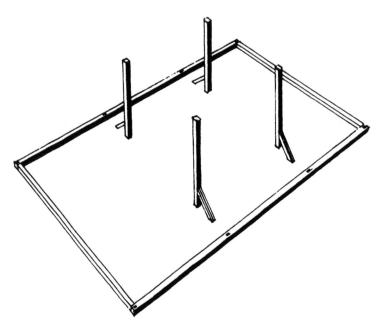

Fig. 29. Belchamp, hypothetical site operation with four posts
reared against earthfast shores.

Finally there seems a possibility that the re-used timbers in the belfry at West Bergholt may have come from a Late Saxon west work, and if this were to be substantiated a case could be made for the development of notched lap joints in England before the Norman occupation. Squinted lap joints exist at Sompting that were never intended to resist withdrawal, but the Bergholt timbers were obviously corner posts from a tower structure posing problems not present at Sompting. It is the minimal degree of resistance to lengthwise withdrawal offered by these examples that states their case for being early and experimental, whilst the resemblance to the angle posts of the Navestock belfry indicates an origin for that type that may be native.

Chapter Two

Examples from the Norman Period (*c.* 1050 to *c.* 1150)

The Church of St. Martin of Tours, Chipping Ongar, Essex

A remarkably complete church in the Romanesque style, and dedicated, it is believed, during the closing decades of the 11th century (R.C.H.M., 1926, Vol. II, 52). Its interest, for the present purpose, lies in the chancel roof which comprises an original area, two subsequent rebuilds and a dated 17th-century reinforcement for the whole. The original part extends seven couples east of the chancel arch and its survival is explained by the stability of the masonry carcase at that point, the chancel walls being buttressed by the cross walls either side of the chancel arch. One of these couples is drawn as Fig. 30. All seven couples are the same, forming a seven-canted void beneath their collars, with vee struts above it; the soulaces and and ashlar pieces forming these cants are fitted by notched abutments, and face-pegged, two pegs per joint. This assembly was designed as an arch and exerts a great deal of outward thrust on its imposts, which were in this case adequate to resist such thrust, largely due to the buttress function of the cross walls. The walls were finished in a level string course, and all the couples stand on base triangles formed by sole pieces as illustrated. The vee struts are fitted by barefaced lap dovetails, as is usual in their case. These couples are a survival of the greatest rarity known to the writer; comparable examples, however, are known in Germany and France (C. A. Hewett, 1979, 98; C. A. Hewett, 1974, 16). There is in these examples no penetration of one timber by another, no overlapping of components and no unwithdrawable forms of assembly insofar as the seven-canted archivolts are concerned; the design resists only compression as does that at Sompting, where face pegging has already been described.

The Church of St. Mary Magdalene, East Ham, Essex

This is an essentially 12th-century church with a later western tower. It was ascribed to the first half of the 12th century by the Royal Commission (R.C.H.M., 1921, 59), and a date *c.* 1130/50 seems probable. Pilasters exist both inside and outside the apsidal eastern end which, it has been suggested, may indicate an original intention to vault the chancel (Sir N. Pevsner, 1956, 149). The timber roof to this chancel is among the rarest in the county, and possibly in the kingdom (Fig. 31). It comprises five couples of rafters united by a ridge piece, with collars and mountants, plus 13 radially disposed rafters over the apse. Thatching battens survive, at unequal heights. Fig. 23 shows the method of laying the sole pieces round the curved top of the apse wall, evidently from east to west, with distance pieces of wall plate between them; 14 short lengths of curved timber were necessary for this purpose. It is inexplicable that this roof has survived. The reason must be a light cladding such as thatch during the greater part of its life and the support given by the cross wall of the next cell of the church, to the west.

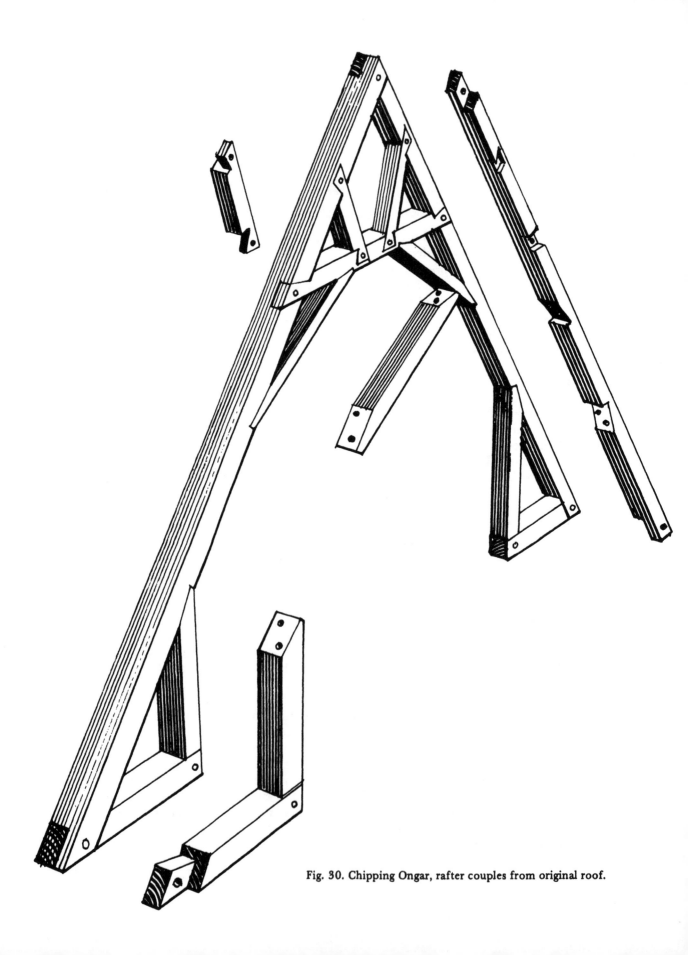

Fig. 30. Chipping Ongar, rafter couples from original roof.

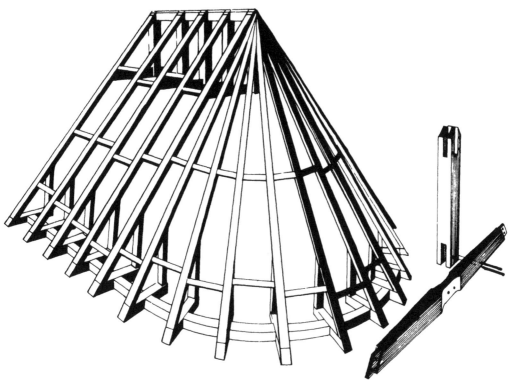

Fig. 31. The apse roof at East Ham with collar and king post shown exploded.

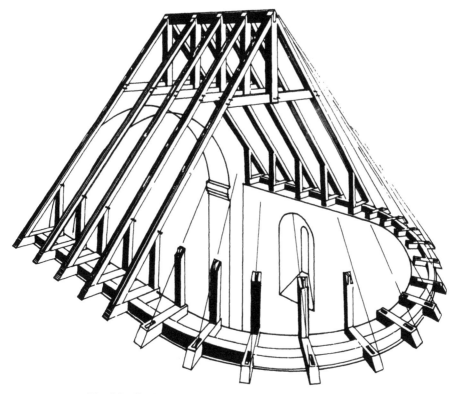

Fig. 32. The East Ham apse roof with sole pieces in place.

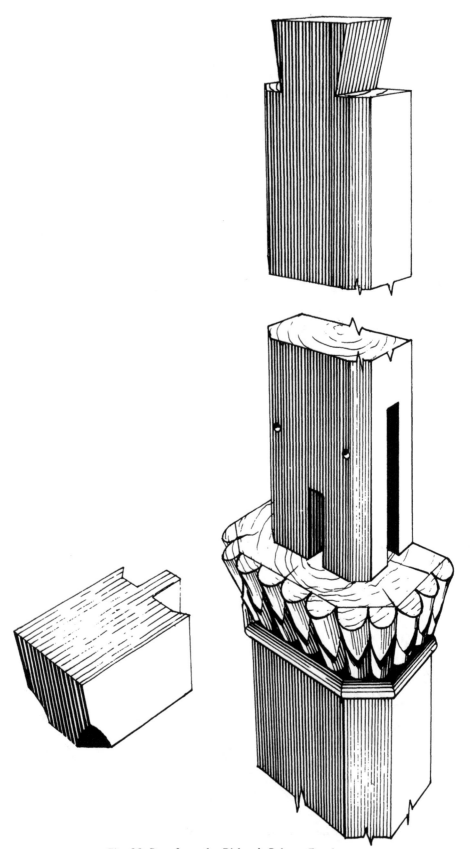

Fig. 33. Post from the Bishop's Palace, Farnham.

The Bishop's Palace, Farnham, Surrey

Some, at least, of the main posts of a great hall survive here, together with an interesting roof of an evidently later date. The one post examined is illustrated as Fig. 33. This has a decorative capital which is integral with the post, and the empty sockets give some indication as to the framing of the original building, which seems to have had its arcade plates fitted in the manner of lintels—without recourse to scarfing. The timber corbel on the frontal face may have been involved with the transverse arch braces, which is another matter for further examination. The importance of this example is twofold. Firstly, on the grounds of styling it is indubitably Norman; secondly, it is of great size. The designer sought to build on a grand scale that was either beyond the limitations that the available timber-trees would have imposed, or at the limit of the carpenters' technical ability to frame.

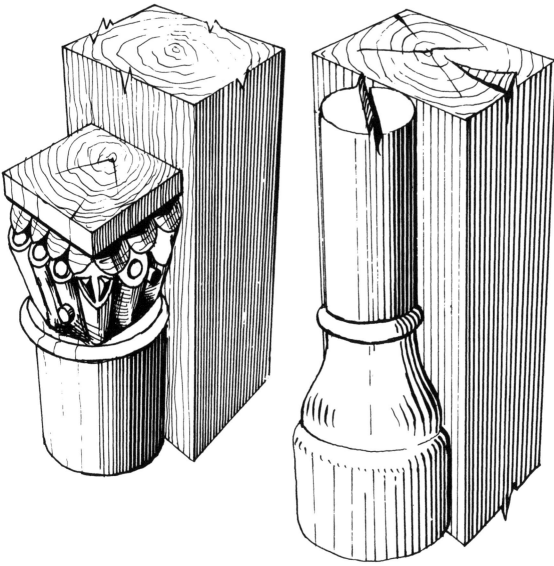

Fig. 34. Post and attached shaft, the Bishop's Palace, Hereford.

39

The Bishop's Palace, Hereford

This palace is today the result of many successive building operations, the most visible of which is that of Bishop Bisse, between 1713 and 1721, who encased the whole building in brick and formed the present hall within it. Rising internally through and above the 18th-century ceilings are the posts and arcades of a huge timber aisled hall, of Romanesque style throughout, and probably the most splendid and impressive example of its kind in England. The quality of craftsmanship is throughout what one would expect to result from prelatical patronage, but because of the great size the jointing of components is often weak and faulty. This has resulted, for example, in the transverse arch braces (which were enormous) being merely butted against the posts and face-pegged, their feet resting on the capitals of the attached shafts, which are separate units of timber. One capital is drawn as Fig. 34, its pegging indicating the use of a free tenon. The same illustration shows the bulbous base of one attached shaft. Fig. 35 shows the evidence that has survived for the mounting of the top or arcade plates of this hall; as in the previous example these timbers served the function of lintels like those of Stonehenge, with which they have more in common than with medieval carpenters' top plates.

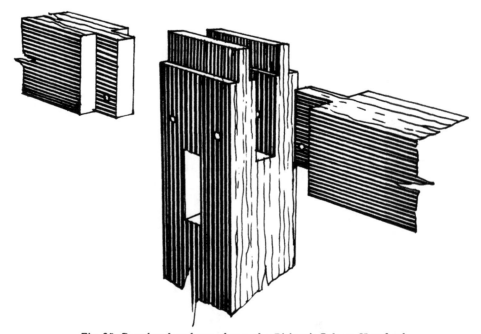

Fig. 35. Post head and top plates, the Bishop's Palace, Hereford.

The assembly method used to fit the great arched braces along the arcades is shown as Fig. 36, in elevation and perspective. This is poor carpentry, and its implications will be discussed when the period is summarised.

40

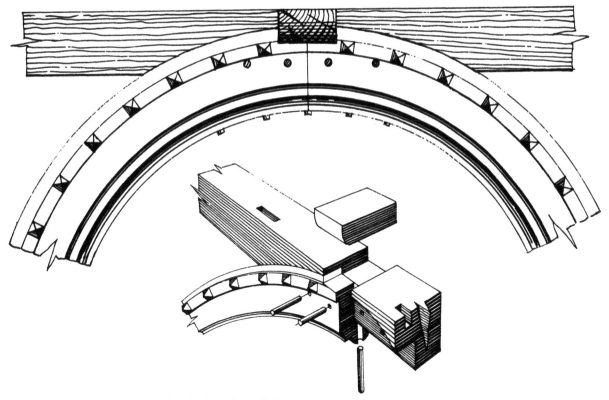

Fig. 36. Jointing of arcade braces at the Bishop's Palace, Hereford.

The Church of St. Peter and St. Paul, West Mersea, Essex

The western tower of this church contains carpentered floors of interest. These originally had the ends of their joists embedded in the masonry of the walls, which is of great thickness; it is undisturbed at the points of entry and the carpentry is assumed contemporary until proved otherwise. The Royal Commission said of it: 'The West Tower (14½ft. by 13½ft.) is of late 11th-century date and of three stages, with an embattled parapet; the quoins are of Roman bricks and the rubble is of coursed ragstone and septaria mostly set herring-bone-wise. The semi-circular tower arch has plain jambs and imposts formed on three oversailing courses. In both the N. and S. walls are late 11th-century windows with round heads and narrow splays' (R.C.H.M., 1922, 231). They did not ascend the tower, evidently, but the latter also has two round-headed windows in the upper stage eastern wall, one on either side of the central 14th-century window and visible only from inside. Whether this tower dates from before or after the Conquest cannot be deduced from the fabric, but late 11th-century (Saxo-Norman overlap) is a satisfactory ascription.

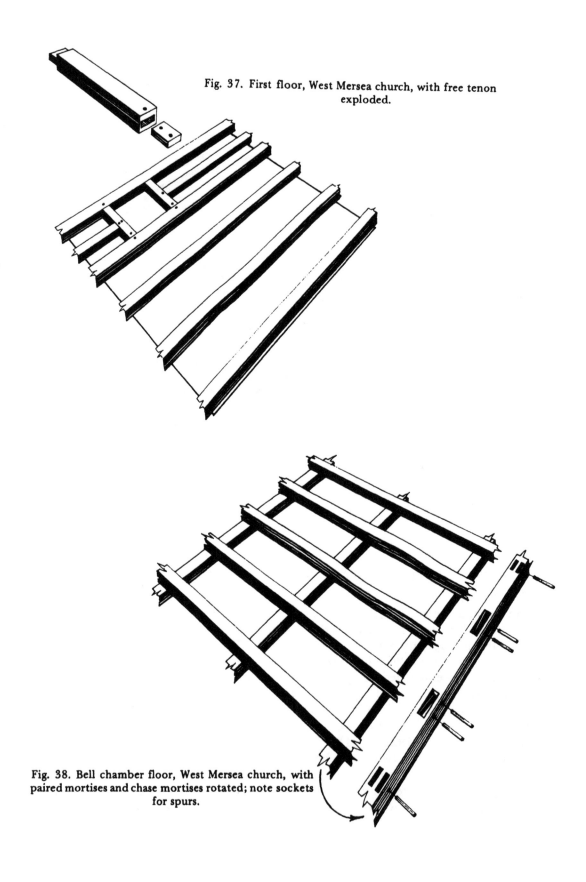

Fig. 37. First floor, West Mersea church, with free tenon
exploded.

Fig. 38. Bell chamber floor, West Mersea church, with
paired mortises and chase mortises rotated; note sockets
for spurs.

The first timber floor comprises five heavy joists aligned from north to south with their ends apparently embedded in the walls, some showing saw marks, others not; between the eastern pair are two trimming joists which frame a bell trap and support two trimmed joists (Fig. 37). The western ends of the trimmers show double pegging, which can only be interpreted as the use of free tenons, on end-grain timber—subject to opening the joints for proof. The second floor was designed to carry the bell frames and is much more heavily framed (Fig. 38). Three heavy joists run north to south, again with embedded ends, and have five almost equally heavy joists laid across them, also with embedded ends. The first three, especially those at east and west, show the joints for wall posts and braces thereto, the impressions of which survive in the masonry, and the jointing of these is advanced for this period. The wall posts each had pairs of single tenons, and their braces had single chase tenons with spurred shoulders, as shown on the overturned joist to the right of the drawing.

Faulkner's Hall, Good Easter, Essex

A complex range of farm buildings here includes a building locally known as the 'Maltings', which purpose it may once have served; it constitutes all that is apparently left of the former prebendal hall. The parish was 'of old called the Prebend of Good Easter . . . because wholly appropriated to the college of St. Martin' (St. Martin le Grand, London). 'Four prebendaries had their corps or endowments in this parish; named Paslowes, Imbers, Fawkeners, on the south side of the church . . .' (P. Morant, 1768, 458; *see also V.C.H.*, London I, 558). These were prebendal manors that may date in their enduring form from 1158, when the endowments of St. Martin were reorganised and the connection of the collegiate foundation with the Honour of Boulogne to some extent broken (*V.C.H.*, London I, 558). Morant's south position was erroneous, but that there is no mistake of identity is guaranteed by a coloured estate map of Samuel Walker's dated 1623 (E.R.O. D/D V28/60), showing the existing buildings almost as they stand today. Evidence for the continued association with the college is available for the period 1379-81 (A. K. McHardy, 1977, 4), in which 'Mr. John Skyrlowe prebendary of Faucons estimated at £12, paying 5s.', this being taxation of the clergy. The fourth prebend was named Bowers, and the buildings were ruinous at the beginning of this century. Nothing now remains.

The main posts of what appears to have been a three-bay aisled hall, framed in heavy oak, survive in the 'Maltings', together with a number of other ancient timbers exhibiting curious assembly techniques that are not apparently in their original contexts. The posts have carved capitals of Romanesque style, with a main span of 17ft., but the spans of the former aisles cannot be deduced. One post drawn to scale constitutes Fig. 39, and Fig. 40 illustrates the details of the capital. The important feature is the accurately cut carinate fillet under the abacus, which derives from a widespread and persistent usage in stonework; but here if considered with the abacus, it illustrates a carpenter's interpretation or idiom, not precisely paralleled in any known early hall. The curvature of the arcade braces is slight, being a quarter of an inch in a length of 8ft.; many posts are trenched for passing braces, none of which have survived. Some posts also have upstands *behind* the top plate, and these may be original features, but the design of the roof has not been

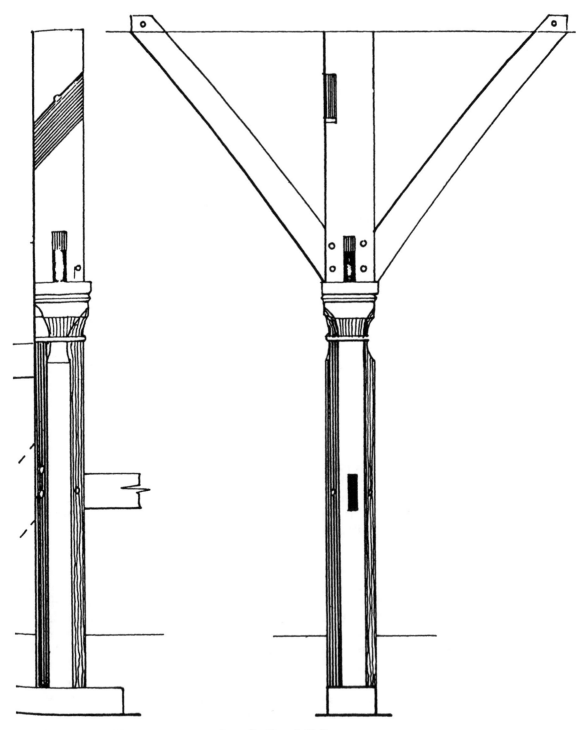

Fig. 39. Scale-drawn post from Faulkner's Hall.

44

deduced because the evidence seems insufficient. The rear faces of all the posts have large chase mortises near their bases for the use of earthfast shores during the rearing process, and their feet are tenoned into short sill pads; there was no connection with the outer walls at ground level.

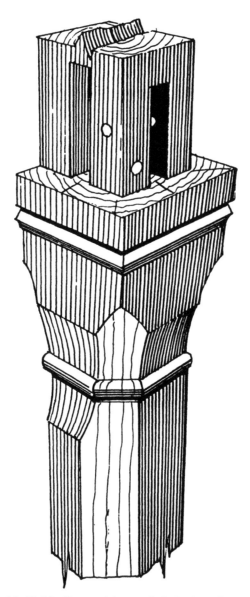

Fig. 40. 'Cubical' or cushion capital, Faulkner's Hall.

The Parish Church, Kempley, Gloucester

The guidebook to this building (Canon D. Gethyn-Jones, 1957) does not give the dedication. The book states that Walter de Lacy of Lassy in Normandy was granted extensive lands, which included Kempley, by William I in recognition of military services rendered. Walter de Lacy died in 1085 from a fall from the battlements of a church he had built in Hereford, and was buried in the chapter house of what is today Gloucester Cathedral. Walter's son, Roger, inherited and extended the de Lacy estates but was exiled for rebellion in 1095, when his estates passed to his brother Hugh who was a devout churchman and was associated with the building of Llanthony Abbey. Between 1090 and 1100, the guidebook suggests, Hugh de Lacy took down the Saxon church at Kempley (for the former existence of which evidence has been found), and this same Hugh died in 1121.

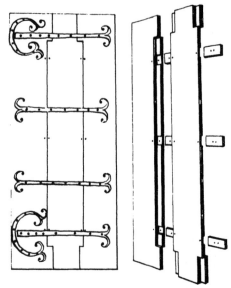

Fig. 41. West door leaf from Kempley church, exploded at right.

These few facts do not clarify the dating of the nave, which in its western end has a Norman window and at ground level a contemporary doorway with door leaf and ironwork, both of which are now addressed to the interior of the west tower, itself of early date. This last fact accounts for the survival, in unweathered condition, of the oak door leaf which is illustrated in Fig. 41. This is remarkable in that it comprises only three planks which are counter-rebated together in the simplest manner, two returns per plank, a fact that allows of a construction date soon after the inception of this structural principle (compare with, for example, Castle Hedingham or Ely Cathedral). More remarkable than this simplicity, applied as it is to an elaborate principle, is the use of free tenons as shown in the exploded drawing to the right of the Figure. This is rare, not being known in other doors, and was certainly an unnecessary additional technique which was omitted from the later doors described, which employed more prolific rebating. It is possible in the light of these peculiarities of construction that the door was constructed at a date close to c. 1100, in Hugh de Lacy's times, when a sufficiently expert craftsman would have been available from Gloucester—with whose Abbot, Serlo, de Lacy had formed a friendship.

The Chapel, Harlowbury, Harlow, Essex

According to Morant (P. Morant, 1768, 483) this manor was given 'to the abbey of Saint Edmund's -bury, by Thurstan, son of Wina a noble Saxon, in Edward the Confessor's reign'. He also states that: 'The Abbot held this maner of the King, *in capite*, as parcel of his barony'. The abbot's great hall largely survives within the present stock brick house; it is an aisled timber building of the early 13th century, with chapel standing close by and south-west of the house. This is a late chapel with datable Norman features such as the handsome north doorway of two orders with attached shafts, on which grounds the whole has long been ascribed to *c.* 1180. At some uncertain date close to *c.* 1300 it was re-roofed with a crown-posted timber roof of considerable interest, the latter having for a long time attracted more attention than the substructure upon which it was superimposed. The rubble 'masonry' of the carcase is conspicuous in that very few of its courses are either level or straight, the single exception being a course of freestone lumps—not ashlar—that once denoted the eaves line. Above this course the walls pitch inward or 'batter' at 57deg. to the vertical and display 17 inclined trenches, which were packed with rubble and mortar when the roof was replaced (Fig. 42). Fig. 43 shows diagrammatically and in perspective how the later roof was placed on the earlier carcase work. The drawing shows the earlier, and probably original, gable couple fitted into a built trench in the rubble wall, and an example of the later couples which were differently jointed at all points; both couples and the collar purlin are shown sufficiently raised to expose their seatings.

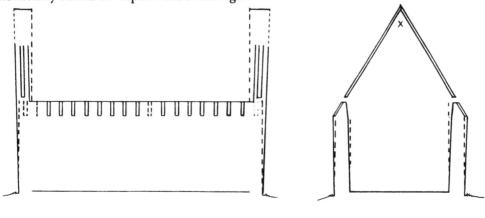

Fig. 42. Scale-drawn sections along and across Harlowbury chapel.

The Barn, Grange Farm, Coggeshall, Essex

At the time of writing this consists merely of three bays, in a dangerous state of ruin; it was formerly a large barn six bays long with two terminal outshuts, or *culatia*. It is situated half a mile from its proprietor, the Savignac abbey of Coggeshall, a royal foundation of *c.* 1140, which was forced into the Cistercian connection in 1147. The proximity suggests a slightly remote *curia* of the main settlement rather than a typical detached Cistercian grange, and 1140 is a valid *terminus post quem* for its building. The foot of one north-eastern post has been carbon-dated by Harwell to 1020 ± 90 (Harwell, 1976, HAR-1258).

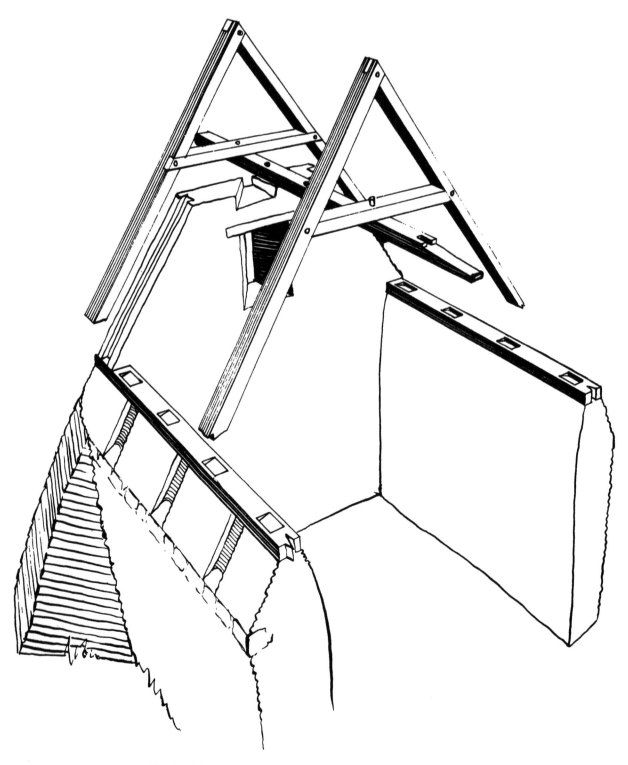

Fig. 43. Original and subsequent roof-mounting at Harlowbury.

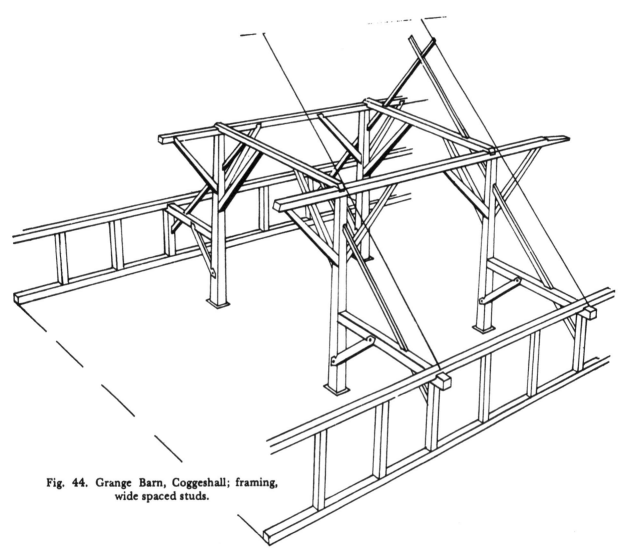

Fig. 44. Grange Barn, Coggeshall; framing,
wide spaced studs.

The main posts stood upon stone stylobates about an inch larger all round than the posts' feet. Examination showed that some stylobates were made of a hard mortar-like mixture, indicating a shortage of the natural stone. Of the latter Mr. Martyn Owen of the Geological Museum, London, reported: 'The specimen is a rather brownish, crystalline shelly limestone and matches well with the specimens in our collection from the Middle Purbeck of Dorset. It is not the Purbeck Marble, which is the upper Purbeck, but from the underlying strata which have been extensively used as building stone'. By the 1170s the distribution from Purbeck quarries was widespread, and the royal interest may have facilitated it. The bases of fallen posts can be seen, and these appear to be shot with a plane to a perfect surface, with no traces of jointing or registration upon their stylobates. The frame of the barn, insofar as its original design can be deduced from what survives, is shown as Fig. 44, and the base of a single post as Fig. 45. This is a very early example of a timber building so designed as to be independent of its site, and was reared by some newly-devised method (concerning which no evidence has been detected), with no recourse to earthfast shores, but it lacked the basal stability afforded by such ground silling systems as were to be evolved later.

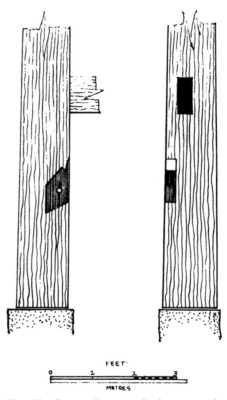

Fig. 45. Grange Barn, scale-drawn post's base; note 'square' mortise and refined entry to notched lap.

Waltham Abbey, Essex

The surviving western part of the former Abbey is the present parish church; two-thirds of its original eastern structure has gone. It was founded in 1030 as a collegiate church of secular canons and refounded as an abbey of Augustinian Canons in 1177. The Norman nave that survives can be dated closely by the pier-base profiles to the decade 1120-30 (S. E. Rigold, 1978, 117; and Fig. 92), and the masonry is of a single build from east to west and from floor to eaves. The internal ashlar is of Reigate and the external of Caen. It is apparent that the church was initially parapeted and later converted to eaves. It may be that until 1807, when the present roof was made out of an earlier one, much of the original roof had survived, and that this was what was re-used at that time. This can be reconstructed from measured drawings of the components; the northern wall plate survives *in situ* at the west end, and a portion of the eaves plate has also been re-used in that area in two bays at the north-west. The reconstructed roof is shown in Fig. 46. The tie beams were placed across the feet of every couple and the couples were widely spaced at centres of a yard, each couple having two collars and soulaces, with notched lap joints at all points. The ashlar pieces were lap-jointed without any notching in order to resist compression. The tie beams were cross-cogged over the partly embedded wall plates and tenoned into the eaves plates—a device not known in any other context. The rafters, all of which were common, were seated in housings on the upper faces of the eaves plates.

50

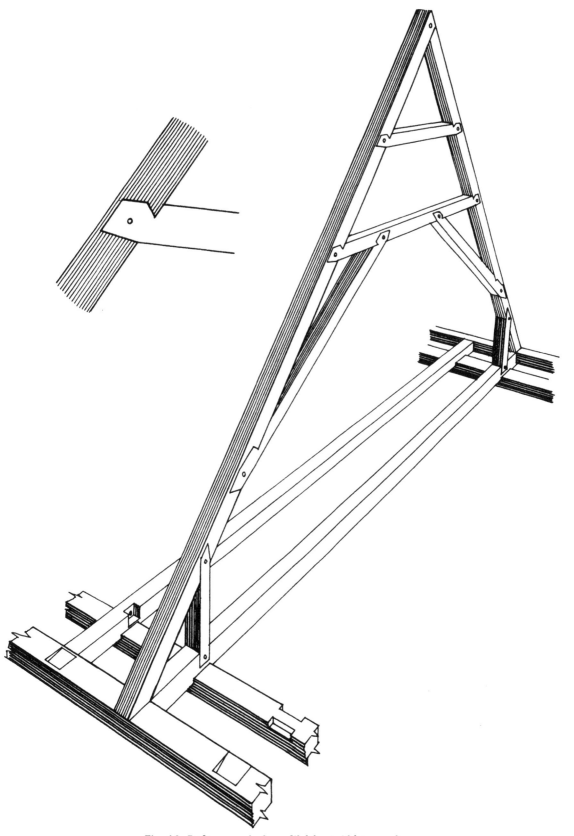

Fig. 46. Rafter couple from Waltham Abbey roof.

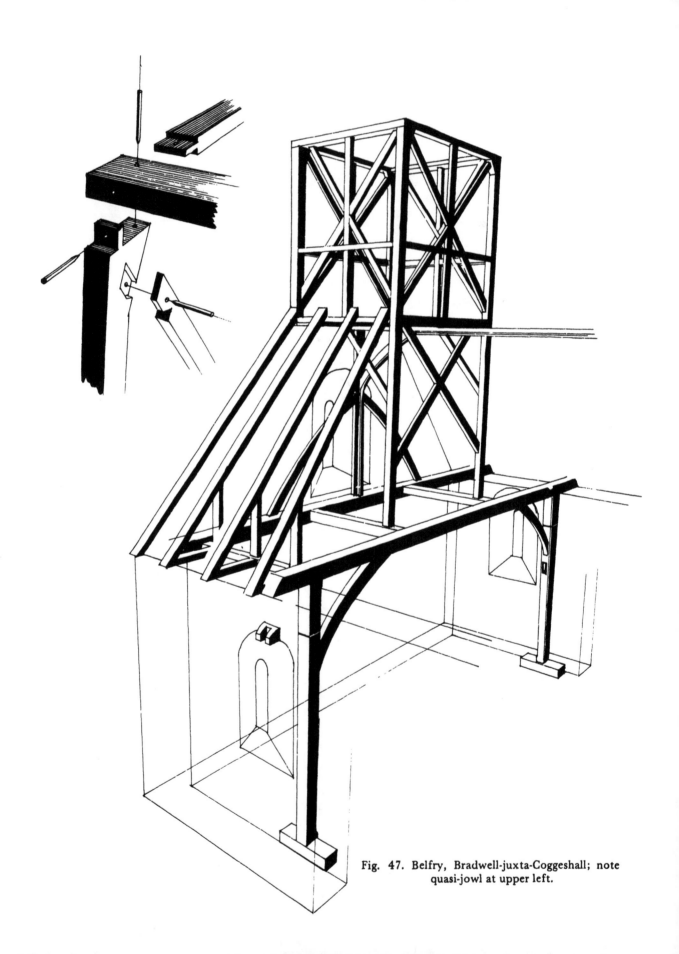

Fig. 47. Belfry, Bradwell-juxta-Coggeshall; note quasi-jowl at upper left.

The Church of the Holy Trinity, Bradwell-juxta-Coggeshall, Essex

One of the most important and instructive churches in the county and one that is essentially Norman in all parts, saving the south porch, which is 14th century and later. Architecturally the important point is that it establishes that the design —a nave with a slightly lower and narrower chancel and western-gable bell turret— is Norman, and not, as has long been generally assumed, the result of subsequent additions to the basic nave unit. The carcase is of stone rubble with large quantities of indurated gravel, probably of local extraction, and Roman brick trim. The murals, 'though not well preserved, make the church one of the essential ones to visit in Essex. They date from about 1320 and are aesthetically of the highest quality, not at all rustic as so much English fresco-work' (Sir. N. Pevsner, 1956, 85).

The frescoes help to date the raising of the nave walls since it was done before the paintings, probably during the late 13th century—a fact that helps to determine a sequence of events that affected the carpentry, and postulates early dates for certain features. When the eaves of the nave were raised the portal frame, on which the eastern bell turret stood, became too low and had to be heightened to restore the relationship of roof and turret; a diagrammatic view of the west end with the belfry is given as Fig. 47. All the evidence at Bradwell is very complete, and the 12th-century transom of the portal frame was simply sawn through, flush with the inside wall, and its outer ends with sockets for rafters were left *in situ* as illustrated. Two short lengths of posts were then added to increase the posts' heights, and the transom was laid on top. The turret was presumably raised at the same time. The latter is assembled with 'lace timbers', long slender braces designed to resist extension, which form saltires on each wall face and terminate in notched lap joints. The nave roof was seven-canted and also had to be raised at this time, when crown posts were intruded simultaneously with its reconstruction at the higher level.

Fyfield Hall, Fyfield, Essex

This is today a capital manor-house, of complex plan and comprising many periods of building; it is situated close to the River Roding and a little north of the church, at the village centre. According to Morant (P. Morant, 1768, 134): 'In King Henry the Second's reign, Pharin de Bologne, and afterwards William de Fessues, or Fésnes, held, of the honor of Bologne, 6 knight's-fees; and among the rest, Fishid. At the same time, Fifid is mentioned among the fees of Oger de Curcun.

'Very soon after, it was in the family surnamed De Tani, lords of Stapleford, and several other estates in the county; namely, in Hasculf de Tani, and his wife Maud.— Then in Rainald; and his brother Grailand, or Gruel de Tani. This last dying in 1179, was succeeded by Hasculf; and he by Gilbert de Tani; who, at the time of his decease, in 1220, left William de Fauburgh; Maud, wife of Adam de Legh; and Nicholas de Beauchamp, his next heirs'. From 1221 it was divided between co-heirs, Langton and Beauchamp who, however, seem to have arranged at various times that one or other should enjoy most of the manor. Thus Stephen de Langton seems to have reunited most of it three years before he died in 1261, and Roger de Beauchamp seems to have held the whole in chief at his death in 1281 (*V.C.H.*, 1956, 46-7). These bewildering details of possession serve to illustrate the fact that some very substantial dwelling-house, probably on the existing site, had good cause to

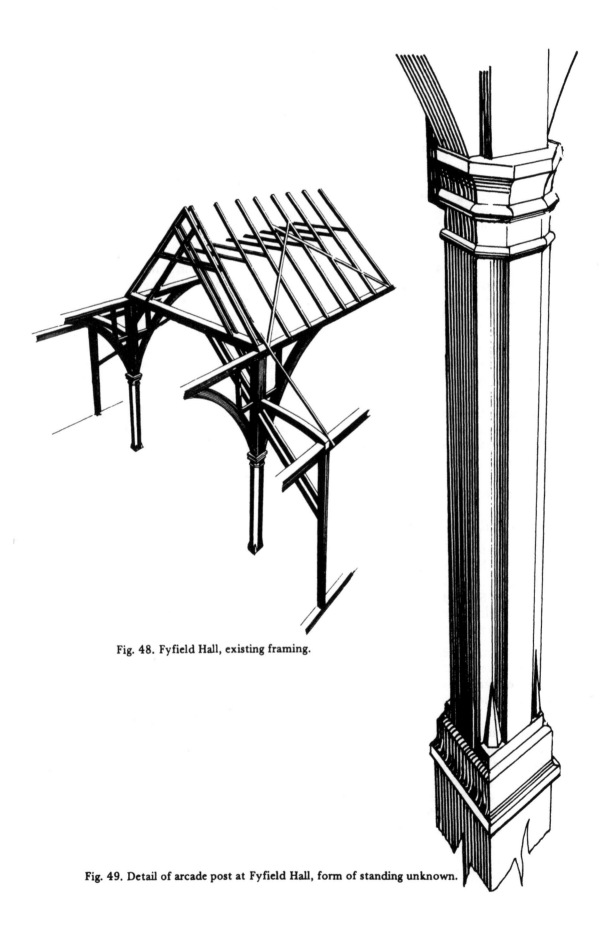

Fig. 48. Fyfield Hall, existing framing.

Fig. 49. Detail of arcade post at Fyfield Hall, form of standing unknown.

exist from as early as 1154; and also that the reunification of the manor, *c.* 1258, might have been sufficient reason to explain an extensive rebuilding of the house.

The surviving two-bay aisled hall does indicate a major rebuild in that its main-span roof and tie beams are not integrated with the substructure. This roof with its tie beams is datable to the 13th century by means of several comparisons which will be made later, whilst the aisled substructure readily admits of an earlier dating. For the present purpose this base-build, comprising six free-standing posts with longitudinal arcades, and its outshut roof (which survives at the north, and possibly also the south) is of greatest interest and the existing roof will be described later, when it seems chronologically appropriate.

The posts themselves, of which the north central one is most intact and exposed to view, are of massive dimensions. They are cut beneath their arch braces into octagonal sections, the inner facets of which are hollowed, and the outer facets virtually flat, although some hint of convexity seems apparent; their capitals present a square abacus with an approach to the beaked half-roll beneath a deep hollow, and a plain roll for the astragal. According to Forrester (H. Forrester, 1972, 31) the astragal is appropriate to between *c.* 1160 and *c.* 1240, and the beaked half-roll between *c.* 1210 and *c.* 1250. In this respect the half-roll at Fyfield is so distinctly two-centred as possibly to be an incipient form. The bases of the posts are more remarkable, having brooch stops to their inner hollow chamfers which form semi-hexagonal cones; these are placed above three outsets, one slightly convex, the next hollow and the lowest distinctly convex. In this way the square-sectioned bases of the posts increase until the latter pass through and beneath the existing floor level—no chase mortises for earthfast shores exist and the method of their basal standing is currently unknown, but sill pads are a possibility.

The arcade braces reach widely and are curved in conformity with the Roman-esque style; they are strutted within their void spandrels by a level and a vertical strut, the second being fitted after the first and into the same chase-mortise. The aisle ties are curved, and in that form they descend to meet the eaves top plates; the outshut rafters are not fitted to the present arcade top plate but ascend a good third of the height of the main-span rafters, where they are affixed to them. Neither free nor wall posts have jowls, a fact dating prior to *c.* 1250. A part of this framing is shown as Fig. 48 and a detail of the visible part of a main post as Fig. 49.

The Norman Period—Summary

It has been stated that 'for a generation or more after the Conquest the English labour force was largely composed of Saxons unused to large-scale operations. Their employment produced in many instances the phenomenon of the "Saxon overlap", when buildings obviously erected under Norman auspices show Saxon characteristics in detail. It was not to be expected that with such a mixed force of men, with very few masters of really outstanding ability, the vast campaign of works should be carried out without some collapsing' (J. H. Harvey, 1971, 21). The earliest among the works described support this statement; the 10th-century roof at Chipping Ongar and the early 12th-century roof at East Ham do not, between them, display any great advances in structural design or jointing technique, and both relate comfortably to the pre-Conquest examples. In the absence of a study of historic carpentry in Normandy it is of course impossible to determine what, if anything, English carpentry derived from the Norman invasion. However, the remaining eight examples tend to indicate that structural thinking with regard to timber continued to develop both jointing techniques and systems of framing that were in use before 1066—unless it can be demonstrated that these same 10th-century methods were more or less common to both England and the Norman area of France prior to 1066.

It is thought probable that the butted and notched (notched in the literal sense) rafter couples at Ongar illustrate a technique that was widespread, since examples are recorded in both Germany and France (C. A. Hewett, 1974, 63). The 12th-century example at East Ham appears to relate to no other known roof at the present time; it is plain, evidently competent carpentry which must await the discovery of comparable works elsewhere. The method of spacing its sole pieces does forecast the method to be employed at Exeter Cathedral in the 14th century, but no association can be suggested. Both of these works sort well with the pre-Conquest examples, and may for that reason be considered native (i.e., Anglo-Saxon).

The next two buildings, the Bishops' Palaces at Farnham and Hereford, are quite obviously Norman, and it is important that their indentification as such depends upon their ornament and style. The details of the carved capitals indicate a date range between *c.* 1115 and 1145 in both cases, and the accentuated nail head used at Hereford is of a similar date. The great size of these two buildings is their weakness, since it is evident that the timber was used at the maximum size available, and with little recourse to jointing; this increase in scale should have produced systems of carpenters' jointing and framing that were commensurate, but this was not the case. The techniques used to assemble the two were, apparently, less secure than those used for the Sompting spire long before the two great halls.

The Hereford example is wrought to a high standard, but this quality is expressed only in the skilled cutting of the timber and the degree of 'fit' achieved. As illustrated, the jointing is weak and hardly deserves to be called such: items such as the huge arch braces of the arcades were merely butted and face-pegged, or housed, butted, and pegged in two planes, as at their tops. In both cases the top or arcade plates were used in the manner of lintels, without recourse to scarf-jointing. In the Farnham example this is most clear and the lintels seem to have lodged in the indentations either side of the post heads. This approach to timber building at great size is curious and admits of two explanations: it may either illustrate Norman techniques, as well as scale and enrichments, or it represents the earliest surviving achievements of native carpenters, Anglo-Saxons, constructing frames on a scale quite beyond their previous experience. If the latter was the case then this phenomenon suggests that the Normans, who paid for the work, were satisfied and could not import a better technology.

The date of the prebendal hall at Good Easter is uncertain, but it could be as early as the appearance of carinate fillets, which certainly date back as early as the late 11th-century crypt of Worcester Cathedral, wherein examples exist. The frame design of this building relates it to the barn at Belchamp that had origins in a pre-Conquest tradition, and the carinate fillets are the only Romanesque element there. The substitution of a timber sill pad for a lime-cement standing is but a slight advance towards the ultimate groundsill, and the earthfast shore seems still to have been in use at the time this hall was reared, suggesting the continuing use of late Saxon methods well into the Norman period and longer than the time generally allowed for the 'Saxo-Norman Overlap'.

The Kempley door leaf seems to be one of the most experimental of its kind (the counter-rebated variety) known to date, for which reason it is suggested that it may date from the rebuilding of the church by Hugh de Lacy, who died in 1121, but no proof is available for this ascription. It has yet to be determined whether this technique was imported, but it was certainly widely employed during our Norman period. The peculiarities at Kempley that suggest a greater age for this door leaf than most of its kind are twofold: the use of the minimum number of rebates that are effective, and the securing of the resultant leaf by free tenons. It is suggested that the mortises were cut before the rebates, while the edges remained square, to facilitate the process. Free tenons were certainly used at a very early stage in the Mediterranean area, for example by the Romans; they were also used on the London waterfront at the Customs House site (C. A. Hewett, 1975, 115). They were used by the Saxons at Sompting for the Rhenish helm, but their application to counter-rebated doors was superfluous and is unknown in that context elsewhere. Another remarkable detail at Kempley is the use of *iron* pins to transfix the tenons. If a Gloucester Abbey carpenter accounts for this example it would be of interest to know whether Abbott Serlo, de Lacy's friend, employed Norman or Saxon craftsmen.

With the barn at Coggeshall Abbey a major step was taken; the use of stone stylobates was not, in itself, far removed from Saxon lime-cement pads—it was, in fact, the same principle with more durable material, but the discontinuation of earthfast shoring seems to indicate the end of a tradition that had lingered on from the times of the earliest excavated timber buildings in Europe. This barn is the oldest timber-framed design at present recorded that relies on its site for no more than the support of its weight; the secondary timbers, in the form of outshut braces, are addressed upward to the frame itself, and the posts were not connected to their stylobates. The notched lap joints used there, with refined entry angles, are the earliest of that kind yet known.

The roof that was built over what is today the nave at Waltham Abbey was designed and built within our Norman period, but the extent to which it is Norman cannot be demonstrated; comparable examples having tie beams to every couple were published by C.R.M.H. (H. Deneux, 1927), and the origins of the system lie in Belgium, France or Germany rather than in England. Whatever the source of this design with two collars and soulaces with ashlar pieces, it is basic to a long series of developments traceable in English high-roofs. The notched lap jointing at Waltham was used with discrimination, being omitted for the compressive ashlar pieces, but the profiles of the notch laps were not refined of entry as at Coggeshall's abbey barn, and Waltham may predate Coggeshall for that reason. The Bradwell belfry is another work that cannot be dissociated from its Norman context, and it may be that such portal-frame belfries are also of Continental origin, although this remains to be proved. The saltire-bracing pattern of its walls is important, and no earlier examples of its use are known; European influence may thus be demonstrable.

The floors in the West Mersea church tower are probably of that tower's building date, and seem likely to be the work of native craftsmen, both masons and carpenters. The design of the first floor does not follow on from the Colchester (Holy Trinity) example, but the strong floor beneath the bell chamber does in that it has bearing-joists under the common joists, and gains more strength in the form of wall posts and bracing. The jointing in the upper floor may be the oldest known example of the two techniques described—paired single tenons and chase tenons with spurred shoulders. The latter certainly existed at Bradwell, dating from the 12th century, but not the former. The date cannot be proved easily, but until disproved the round-headed windows and herring-boned rubble suggest that it is early Norman. Bracing in these examples is mainly straight, but Romanesque arches exist in the Hereford Palace, and very subtle curvature in the prebendal building at Faulkner's, making it difficult to assess the extent to which the Romanesque style obtained.

Chapter Three

Examples from the Great Transition (*c.* 1150 to *c.* 1200)

The Barley Barn, Cressing Temple, Essex

Accounts of this building containing varying degrees of detail have been published on previous occasions, and these are cited in the references. A summary of the evidence is necessary, however, and the association of the site with the military order of the Knights Templar is the apparent reason for the name of the holding. Of this connection Morant (P. Morant, 1768, 113) says: 'It was anciently possess'd with Witham by Earl Harold, and others, down to the time of K. Stephen; who, about 1150, granted the whole maner of Cressynge with the advowson of the church in the same maner, to the Virgin Mary, and the Brothers of the Knighthood of Solomon's Temple at Jerusalem. Hence it came to be called Cressynge-Temple, as belonging to the Knights-Templar: Who, in 1185, are recorded to have had 5 hides of land in Whitham and Kirsing, of the gift of K. Stephen, one part of which was in demesne, and the other lett to divers tenents; whereof Adam de Kirsing held two virgates at 25s. This was one of the Knights-Templars cells, styled a Commandery, or Preceptory, dependent upon their capital House, The Temple, in London'.

The barley barn is the older of two remaining on the site, and is constructed upon six transverse frames with outshuts, or aisles, on either side, the five bays containing about five hundred square feet each. It has been very much altered during its long life, but enough original evidence survives for a reconstruction to be made. It obviously had terminal outshuts, or *culatia*, before being shortened to its present size, and the method of assembly was by means of chase tenons along its length and open notched lap joints across its width, these last having refined angles of entry. At the time of building the roof was scissor-braced by means of passing braces of great length, which reached from the outshut wall posts up to the principal-rafters in the opposite pitches of the ridged and hip-ended roof. The upper half of the length of one of these very long timbers survives *in situ*.

Perspectives of transverse frames are shown in Fig. 50 (a central frame), and Fig. 51 (a terminal frame). A carbon[14] date was calculated for this building in 1956, as a result of the kind interest of Professors Berger and Horn of California; in the light of six samples of timber from various positions the date suggested was *c.* 1200 ± 60 years, after applying various correcting factors. The use of notched lap joints in this barn in the vertical plane is confined to timbers transversely disposed in relation to the plan—a point of construction which differentiates the barn from the demonstrably earlier barn of Coggeshall Abbey, and also from the framing of Songers at Boxted (which is described later), neither of which are far away.

A recent re-examination has positively identified which of the main posts occupies its original standing; it is the one in the northern corner of the building,

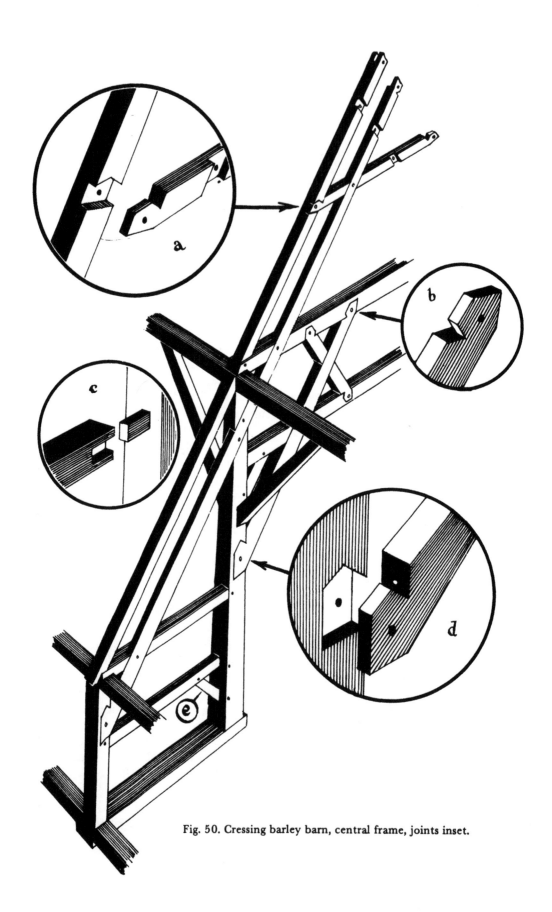

Fig. 50. Cressing barley barn, central frame, joints inset.

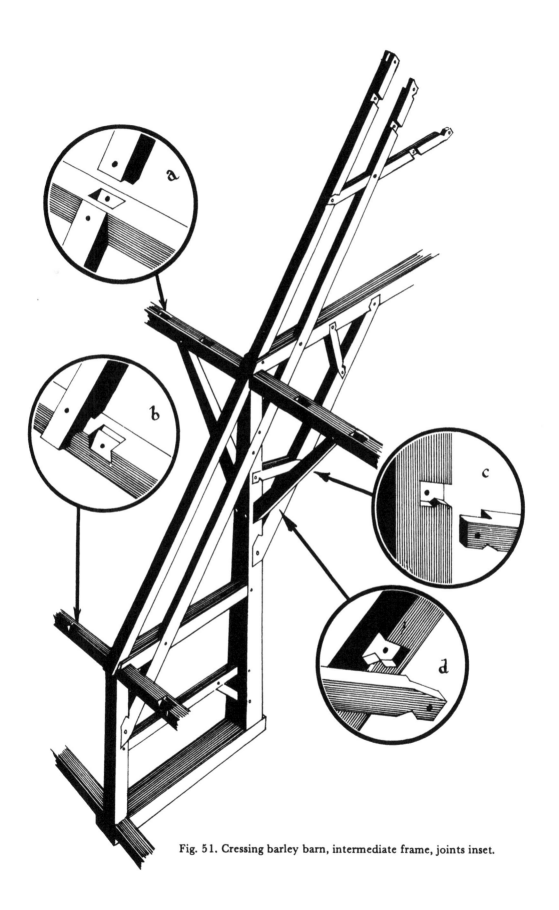

Fig. 51. Cressing barley barn, intermediate frame, joints inset.

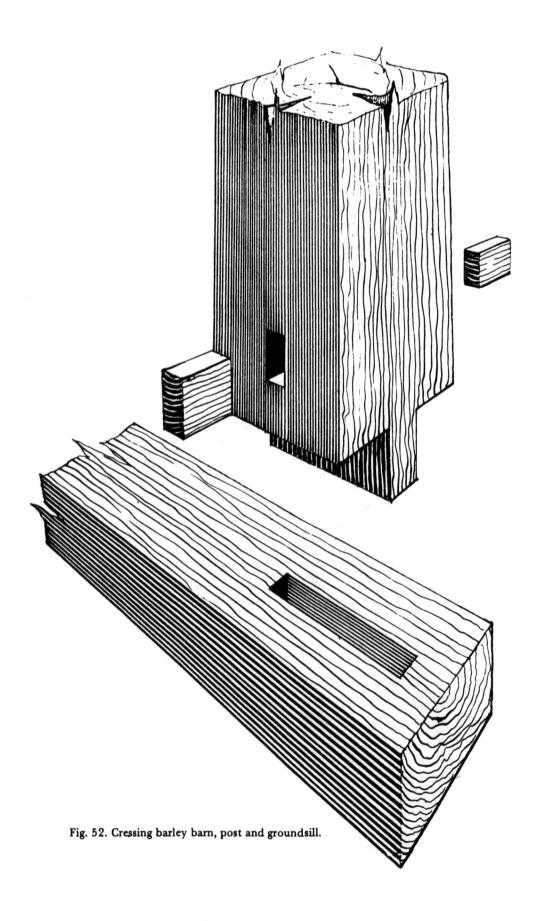

Fig. 52. Cressing barley barn, post and groundsill.

the post that has suffered the most severe depletion of its complementary bracing. This post stands upon its original groundsill, which connects it structurally to the external wall's groundsill, and is illustrated in Fig. 52. It was cut with an integral foot tenon and, like all of its fellows, fitted with side plugs that were tapered both depth- and widthwise to render them withdrawable. The existence of these side plugs took several years to detect since they have exactly the same patination as do the posts themselves, but all 24 are now proven. Their object was evidently to facilitate rearing and to prevent the feet of the posts from sliding when being reared; this postulates a mechanism employing trunnions similar to those of an ordnance carriage, which could be staked firmly to the site for each operation and moved from the first position to that of each subsequent operation. This would have ensured that the foot tenons were accurately registered for entry into the sill mortises, and precluded the possibility of any disastrous sliding inward of the portion of 'arcade' being reared—comprising, when the timber was green, many tons of framing. Manpower would have been inadequate for controlling the feet of such heavy timbers during such an operation as this, and life could have been placed at risk. The taper of the sockets indicated the easy withdrawability of the pivots of the mechanism, and also that the subsequent plugs were merely to conceal the empty sockets. The developmental significance of this procedure lies in the use of the full groundsill, which was integrated into the framing of the exterior walling; and also the fact that all framing within the outshut was here addressed to the superstructures, and upon less experimental grounds than the barn at Coggeshall Abbey Grange demonstrates.

The end frames were fitted with angle ties betwixt the tie beams and top plates, and these have chase tenons at the beams and overlapping notched-lap joints on the plates. This form of the joint (Fig. 53) must be the origin of the secret notched lap, incepted at Wells after the Interdict, the difference being that the latter was flush-fitting.

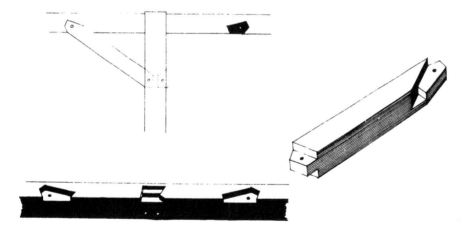

Fig. 53. Cressing barley barn, three views
of angle tie and its joints.

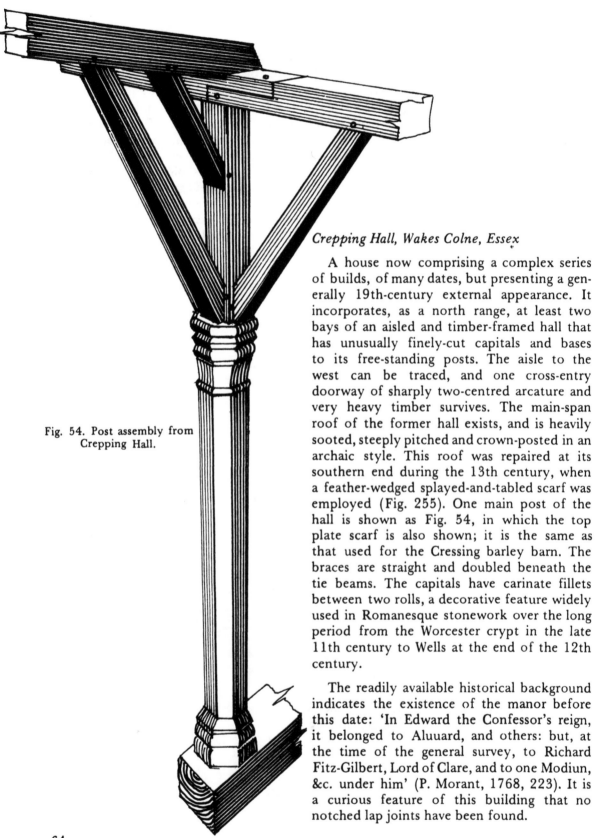

Fig. 54. Post assembly from Crepping Hall.

Crepping Hall, Wakes Colne, Essex

A house now comprising a complex series of builds, of many dates, but presenting a generally 19th-century external appearance. It incorporates, as a north range, at least two bays of an aisled and timber-framed hall that has unusually finely-cut capitals and bases to its free-standing posts. The aisle to the west can be traced, and one cross-entry doorway of sharply two-centred arcature and very heavy timber survives. The main-span roof of the former hall exists, and is heavily sooted, steeply pitched and crown-posted in an archaic style. This roof was repaired at its southern end during the 13th century, when a feather-wedged splayed-and-tabled scarf was employed (Fig. 255). One main post of the hall is shown as Fig. 54, in which the top plate scarf is also shown; it is the same as that used for the Cressing barley barn. The braces are straight and doubled beneath the tie beams. The capitals have carinate fillets between two rolls, a decorative feature widely used in Romanesque stonework over the long period from the Worcester crypt in the late 11th century to Wells at the end of the 12th century.

The readily available historical background indicates the existence of the manor before this date: 'In Edward the Confessor's reign, it belonged to Aluuard, and others: but, at the time of the general survey, to Richard Fitz-Gilbert, Lord of Clare, and to one Modiun, &c. under him' (P. Morant, 1768, 223). It is a curious feature of this building that no notched lap joints have been found.

64

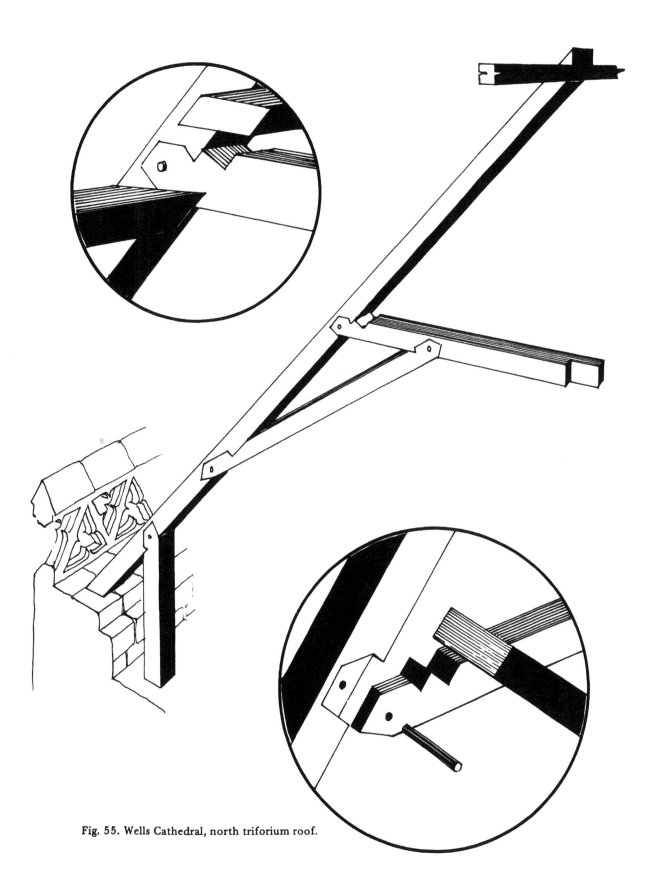

Fig. 55. Wells Cathedral, north triforium roof.

The Triforium Roof, the Nave, Wells Cathedral, Somerset

The earliest surviving lean-to roof in a cathedral triforium appears to be the example illustrated in Fig. 55, which is from the northern triforium at Wells which was possibly completed by *c.* 1180, since the triforium would have needed a roof before the nave arcades had attained their full height. This design has a great deal in common with that of the nave high-roof, but most importantly it was equipped with a side purlin, which could do nothing more than maintain the single pitch of the roof in-plane—i.e., flat. A triforium roof was never likely to rack, since it was located at both top and bottom of its rafters by masonry walls, and the introduction in this context of a member—the side purlin—that was ultimately to be absorbed into ridged roofing for the specific purpose of preventing racking, is interesting. The eaves here was converted into a parapet, as in the nave, so the base triangulation is missing; the span is 16ft. and there is no surviving evidence for tie beams.

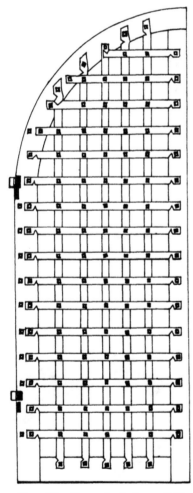

Fig. 56. Door leaf from Selby Abbey, inner face.

The West Doors, Selby Abbey, Yorkshire

According to the guidebook sold therein (Rev. J. A. P. Kent, 1968, 3), the western entrance of this abbey was completed at the latest by 1170, and the pair of oaken door leaves are probably of the same date. These are framed in portcullis fashion with grown bends at the top of their durns; the ledges are applied to both inner and outer faces, sandwiching the planks. The joints around the edges are notched lap joints. The internal face of the left-hand leaf is shown as Fig. 56.

The West Doors, Ely Cathedral, Cambridgeshire

The west transept and tower of Ely Cathedral are dated between *c.* 1174 and *c.* 1197 (J. H. Harvey, 1961, 131), and the door leaf of that date, which was apparently a single one, was built by using the counter-rebated technique. This was cut into a pair of door leaves, each taller, narrower and lancet headed, at the time when the Galilee porch was added, *c.* 1250. The resultant two are shown as Fig. 57, together with an exploded view of the counter rebates.

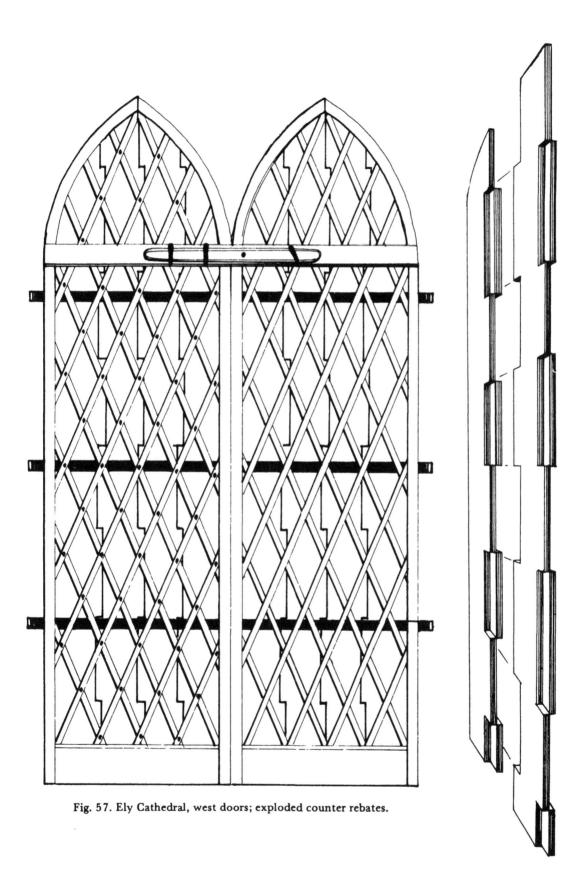

Fig. 57. Ely Cathedral, west doors; exploded counter rebates.

'A complete Late Norman parish church 125 ft. long to the E. arch of a Norman tower replaced by the Tudor Tower . . . Nave and aisles of six bays, and a long chancel. The nave arcades rest on alternatingly circular and octagonal piers with splendidly carved leaf capitals, mostly of crocket-like leaves, but in one case also of real crockets on the French Early Gothic pattern. That dates the nave as not

Fig. 58. Castle Hedingham, south door leaf.

earlier than *c.* 1180. The complex mouldings of the arches indicate so late a date too' (Sir N. Pevsner, 1956, 99). Pertaining to this build and date are the three door leaves of this church, of which the finest, but most insensitively restored, is the great south door (Fig. 58). The important feature of this door, in view of the date quoted, is the liberal use of counter rebating, which is apparently applied without regard to expense to numerous and relatively narrow door planks. The finance for such a lavish exercise in craftsmanship was doubtless provided by the de Veres, Earls of Oxford, and one of the most powerful families in Norman England. At the time of writing no more elaborate example of this jointing technique is known, and the cited date probably represents the high point of such door constructions.

Both the nave high-roof and the triforium lean-to roof at Wells are datable with regard to the 'break'. This is the much-publicised change in ashlar size and certain decorative features that can be seen at varying points along the nave, westward from the crossing. This change can be traced along a line sloping upward and eastward from the fourth bay, west of the crossing, at ground level. This change, it was suggested by Bilson (J. Bilson, 1928, 23–68) may have occurred between 1210 and 1220, and it has since been ascribed to the effects of the Interdict between 1209 and 1312 (L. S. Colchester, 1976, letter to the author). The last quoted date places Wells amongst other works of carpentry that have come to light with regard to this period, and is accepted for the present purpose.

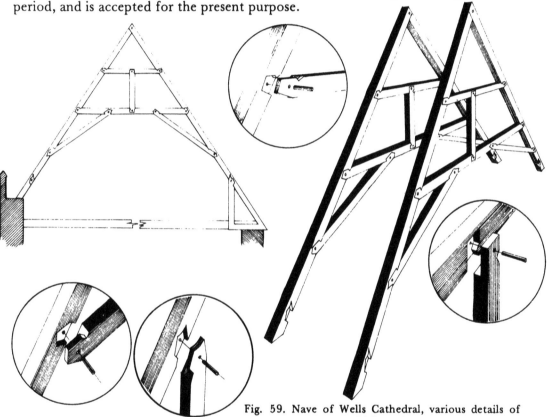

Fig. 59. Nave of Wells Cathedral, various details of high-roof.

The roofs concerned had been built from *c.* 1175 (or possibly a little earlier) until 1209, when work stopped; it was restarted after 1213, and continued until the west front was reached by 1260. Parts, that is some bays, of both triforium and nave high-roofs were thus completed before or by 1209, and both must represent the original design proposed and accepted for the cathedral. The general design is of a base-tied roof having two collars, with mountants between them and in extension, and with soulaces and ashlar pieces—all of which were notched—lap-jointed. This is illustrated in Figure 59, which also shows at upper left the subsequent conversion from eaves to a parapet; part of the triforium roof was shown in Fig. 55, and it relates closely to that of the nave.

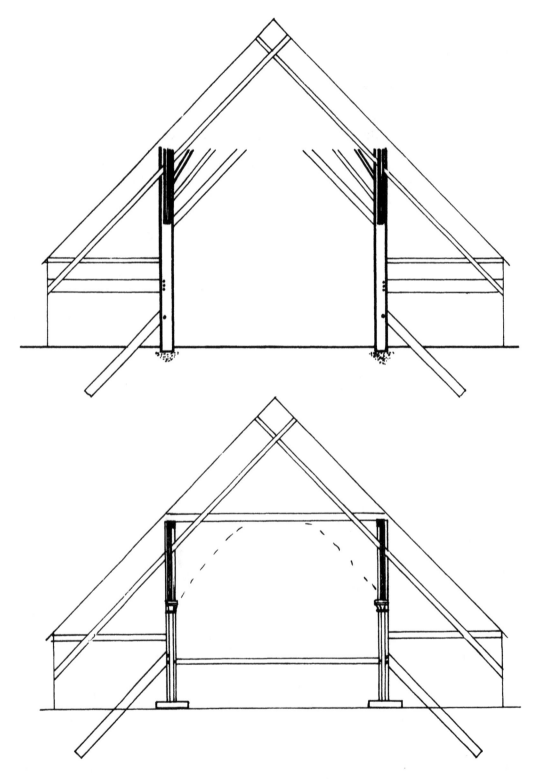

Fig. 60. Four diagrammatical sections, pre- and post Conquest.

70

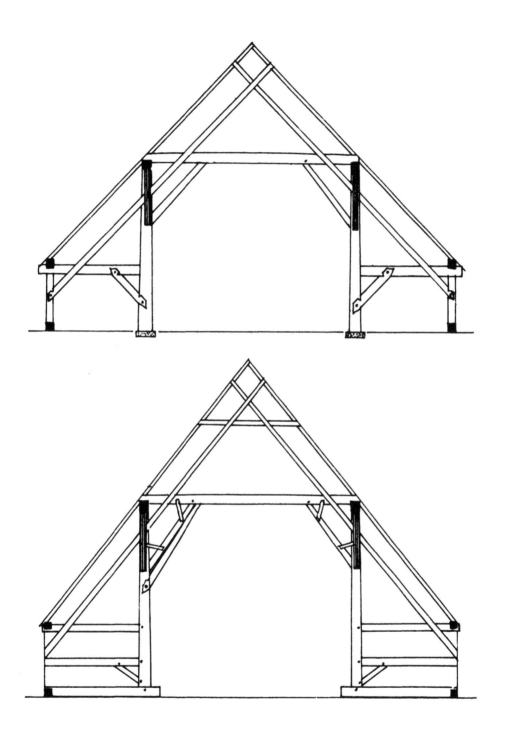

71

The Great Transition from Norman to Early English—Summary

The barley barn at Cressing is not more closely dated than any other building of these times, despite the number of carbon[14] determinations made in connection with it. The problems regarding dating have been previously published (C. A. Hewett, 1969, 56), and for the present purpose the building is regarded as an example of 12th-century carpentry. Since it illustrates what is evidently an early example of a ground silling system for a large barn, concerning the rearing of which there were misgivings, it provides a point from which to review developments leading up to this advance.

From the unknown date of Greensted church, perhaps the mid-9th century, groundsills for external walls, apparently laid directly on the prepared site and without known damp-coursing provisions, can be assumed to have been in use. Before this, and in association also with groundsills, as at Yeavering, external buttresses of timber had been used as described by Bede. It is suggested that the external buttresses were modified prior to and during the 11th century, and were used to facilitate the rearing of free-standing posts in aisled buildings such as the St. Paul's barn, and thereafter the prebendal hall. In both cases heavy straining beams were also used in the aisles, suggesting the use of reversed-assembly at those points. These buttress timbers, or earthfastshores, persisted for a while into the 12th and 13th centuries as sole braces, of which notched and lap-jointed examples exist at Sandonbury, Hertfordshire (another St. Paul's manor), Abbas Hall, Great Cornard, Suffolk, and a chase-tenoned example at Wynter's Armourie in Essex (C. A. Hewett, 1976, 54, 56 and 92). Cross-sectional diagrams of four examples illustrating this transition are given as Fig. 60.

The basal mounting of the posts of Crepping Hall has yet to be determined, but they certainly stand upon timbers of an unknown length, which may be sills or sill pads as at the prebendal hall. As has been mentioned already, both Fyfield and Crepping Halls do not show notched-lap-jointing, except that of later workmanship in the case of Fyfield. The diagrammatic sequence shown as Fig. 60 is also curious in that the least complicated framing was that of the Abbey Grange at Coggeshall; this exercise in economy—for such it evidently was—necessitated the major rebuild, c. 1375, when edged-halved and bridled scarfs were used. The general sequence, however, was the development of the elaborate framing used during the closing decades of the late Saxon period.

The various counter-rebated doors described all date from the last quarter of the 12th century and they not only illustrate the highest point of complexity in this field, but also indicate the approaching end of such work, which must have entailed very high costs. No purely domestic example of counter-rebated door leaves has yet been published, but royal or capital manor houses probably possessed them.

The high-roof of the nave at Wells and the associated parts of the triforium lean-to roofs may be taken as designed and executed well before c. 1200, with particular regard, that is, to the parts immediately west of the crossing. The cross-sectional design of the nave roof derives from that used originally over Waltham Abbey, but it is difficult to understand why the ashlar pieces in this case were fitted with notched-lap-joints. The actual forms of both laps and notches vary, but refined angles of entry were used, as was to be expected, in all cases. No definitive dating of the developments in the form of these joints can yet be deduced, but they will be more fully discussed after what seem to be the ultimate examples have been described. The two high-roofs over the choir transepts of Lincoln Cathedral both seem to be 'departures' for which no comparisons are known to the writer, although some association is evident with Beverley Minster, the nave of which has a roof of a derived type.

The framing of the arcades of Crepping Hall has a lot in common with the barley barn at Cressing; the scarf joint is common to both, and both have straight and square-sectioned bracing that is duplicated under their tie beams, the main differences being cambered tie beams in the house, and an absence of lap joints. The dating of the house must be determined by the capital and base mouldings of its posts, which with regard to the carinate fillet will allow the range 1084-92 until 1295 (J. H. Harvey, 1961, 167; A. W. Clapham, 1964, 123), while the bulbous base moulding also suggests a date within that range. Nothing seems to be known in masonry that is strictly comparable with these bases.

During the time spanned by the seven works mentioned the Early English style of Gothic architecture had begun, apparently at Wells Cathedral, which 'shows us a completely different version of Gothic, devoid of the old Corinthianesque character of Canterbury and markedly English in what has been called the South-Western manner' (J. H. Harvey, 1974, 200). Despite the arrival of this English Gothic style some works in the earlier Norman style were finished according to their original design, and as with developments in jointing it is possible to deduce when a new concept—either structural or decorative—was first applied; but difficult, if in fact possible, to determine when its predecessor went out of use. For this reason no precise chronological order can be established for the seven works in question.

As with previous examples it is difficult and in most cases impossible to relate purely functional carpentry to the prevailing decorative style in architecture, and thereby broadly date it. Few carpenters even at cathedral level applied mouldings, capitals or bases to their works, and the variations in type of arcature were reflected, in timber braces, mainly in buildings where they were a displayed feature. The carpenters responsible for the cathedral high-roofs, and the great barns of the monastic and military orders, were, as is proper, more concerned with the mechanics of their designs and the most appropriate jointing. However, good examples of elaborate and datable moulding, when found in timber buildings such as the Good Easter prebendary, should not be regarded as later copies of preceding fashions in this field: it is considered that in view of their expense they must be of precisely the same date as their masons' counterparts.

The oldest parts of Fyfield Hall cannot be dated other than within the range suggested, between c. 1160 and c. 1240, mainly in the light of the mouldings of the capitals. However, in view of the style of the arcades, which more closely approach round arches than any others listed, it is possible that the date may be earlier than 1200, and within the Norman period.

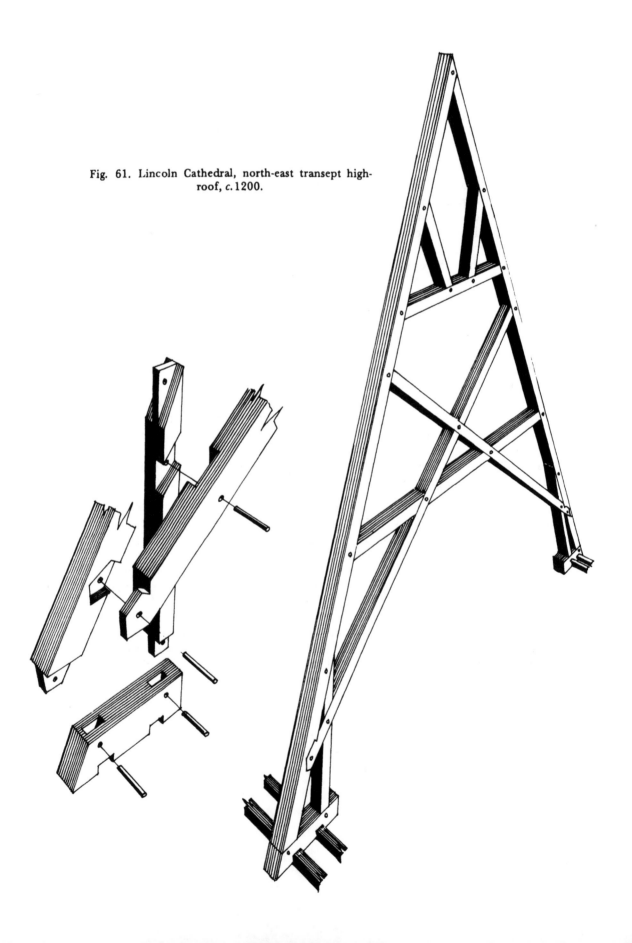

Fig. 61. Lincoln Cathedral, north-east transept high-roof, *c.*1200.

Chapter Four

Examples from the Early English Period (*c.* 1150 to *c.* 1250)

The North-East Transept, Lincoln Cathedral

The choir and east transept at Lincoln were built between 1192 and 1200 (J. H. Harvey, 1961, 141), and the craftsmen responsible are unknown. It is probable that the existing eastern transept's high-roof can also be dated thus, for there is no evidence to the contrary and the design and jointing of the carpentry fits well among similar and contemporary works. This provides a rare example of wall plating here, since most of the Lincoln roofs do not appear to have been similarly fitted; a frame is illustrated in Fig. 61, in which an exploded view of the eaves triangulation is given to the left. Unwithdrawable joints, in the form of notched laps, were here confined to the bases of the scissor braces, which a brief study of the drawing will establish were the only points at which they were appropriate. The bays of this roof were very short, having only three common couples between tie beams, and the profiles of the laps' notches are distinctly archaic.

The South-East Transept, Lincoln Cathedral

The high-roof of this differs substantially from its northern counterpart, although the exact profile of the notched-lap-joints is the same. Wall plates were again used, the internal one of the pair provided each side having a hewn fillet along its upper face. The long and canted ashlars of the northern example were here replaced by truncated under-rafters, scissor braces being dispensed with. The bays were equally short, having the same three couples betwixt tie beams, and the latter were, curiously, overlaid with longitudinal timbers as illustrated in Fig. 62. This design produced what has proved to be an efficient, non-spreading rafter couple, having a high mid-span clearance that the vaults' crowns did not necessitate. It was a design, however, that was certainly used for the older part of the nave high-roof at Beverley Minster, and may have influenced other architects of the time. So far as is known the whole eastern transept dates from between 1192 and 1200, despite the differences between the north and south roof designs.

The Church of St. Mary the Virgin, Syde, Gloucestershire

The parish of Syde contains only 628 acres and claims to be among the smallest and most ancient in the kingdom; its parish church is small and is considered to date mainly from the 12th century, an assessment resting upon the visible features of the masonry. Its main interest lies in the timber nave roof which is of four bays and mounted on relatively high side walls, and framed into seven cants, the joints of which have been wrapped with iron—precluding any examination. Under the roof three archaic crown posts were intruded at some unknown date. This is

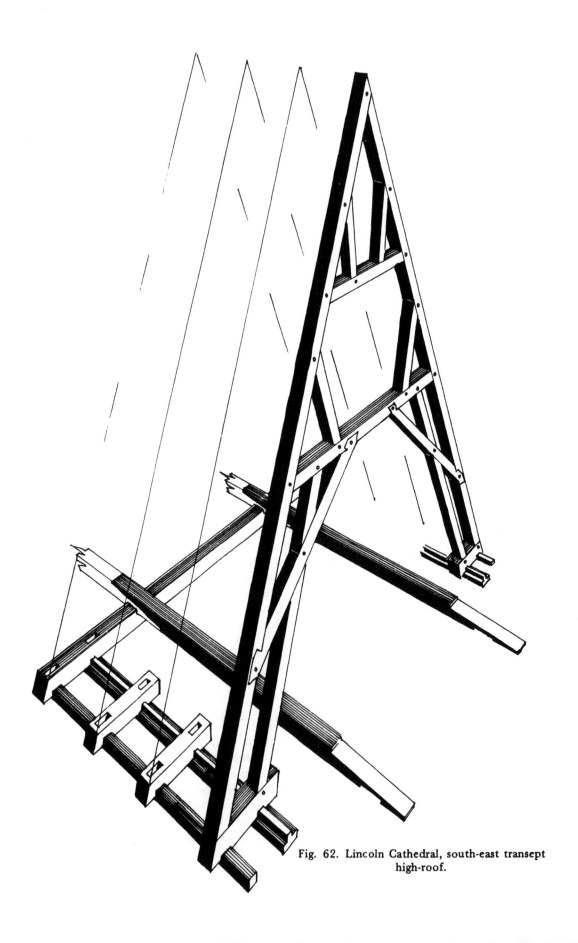

Fig. 62. Lincoln Cathedral, south-east transept
high-roof.

illustrated as Fig. 63, and the structural details as Fig. 64. The lap-jointing of the crown posts to the tie beams' sides (shown at D in Fig. 64) indicates that they were intruded, because crown posts are normally tenoned into the beams' upper faces; these are the only examples known that are assembled by open notched-lap-jointing (as shown at B and C in Fig. 64). The braces are of the earliest type and spring near to the posts' feet, straight, square-sectioned and steeply pitched. The collar purlin is jointed with a through-splayed scarf, a further indication of an early date (as shown at A in Fig. 64).

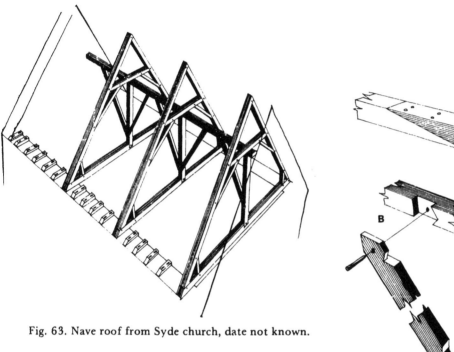

Fig. 63. Nave roof from Syde church, date not known.

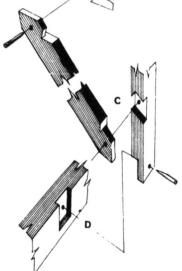

Fig. 64 (right). Structural details from Syde roof.

If, during the 13th century, so small a church went to the expense of using a carpenter of the necessary calibre to use notched-lap-jointing on so small a scale, as seems probable, then his technique was likely to be up-to-date and the work may be of the beginning of the 13th century. It this were proved, the seven-canted earlier roof may be of an earlier date, such as that of the Chipping Ongar example.

The Church of St. Thomas the Apostle, Navestock, Essex

Much has been published concerning the western timber tower of this basically Norman church, and the tower itself was carbon-dated in 1963, giving a date range of 1133 to 1253 (previous publications were C. A. Hewett, 1969 and 1979). The tower was built to the south-west of the same (south-west) corner of the Norman nave, and the south aisle was built in close chronological succession. The building of the aisle joined the tower to the church and provided covered access into its base; the arcade is an example of carpentry faked to represent stone (C. A. Hewett and J. R. Smith, 1972, 84), and it is datable by its mouldings to c. 1250. The exterior aisle walls were of real rubble masonry and prove the building sequence, in that the western wall's masonry protrudes into the frame spaces of the tower.

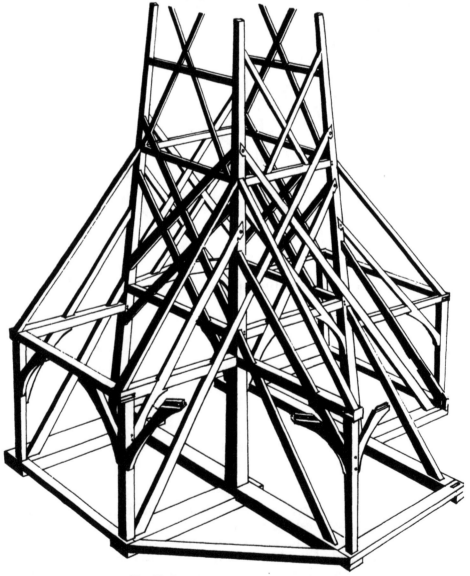

Fig. 65. Navestock belfry, base of framing.

The latter was built with in-canted posts and was elaborately braced (or laced) with very long, thin timbers acting in extension and resisting it by virtue of their notched-lap-jointed ends, the notches in question being of the 'archaic' profile. One of these joints, from the top of a long brace, is shown in Plate XII. It broke in ancient times, as may be seen from the photograph, and exemplifies the type of joint failures that undoubtedly motivated the development of all types of carpenters' joints. The tower framing without the spire constitutes Fig. 65.

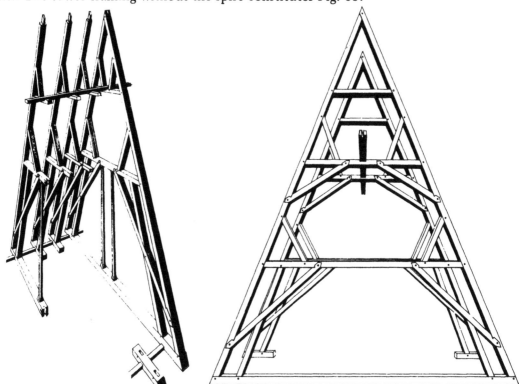

Fig. 66. Lincoln Cathedral, high-roof to nave west of crossing.

Fig. 67. Lincoln Cathedral, high-roof to nave and collar purlin, west end.

Lincoln Cathedral

The nave of Lincoln was roofed by *c.* 1253 when the west front was completed, it is believed under Master Alexander (J. H. Harvey, 1961, 141); and parts of the roof must have been in position by *c.* 1225, at the beginning of the work. There is certainly a development in the design during the progress of the roof from the west face of the central tower toward the west front, and illustrations of the two methods are given as Figs. 66 and 67. Immediately west of the tower, tie beams were fitted to every fourth rafter couple, and very slender and elegantly chamfered posts were set beneath the collars—a design which evolved from the earlier roofs of the eastern transepts.

The open notched lap joint was favoured for the unwithdrawable ends of the very long soulaces, and a collar purlin was introduced to prevent racking; this was housed *over* the collars.

The Church of St. Peter, Roydon, Essex

This church was ascribed to a date between *c.* 1220 and *c.* 1250 (R.C.H.M., 1921, 207), mainly on the basis of a single renewed lancet window on the south side. The true date for its oldest part, the nave (insofar as it survives), cannot be deduced, in view of the lack of visible evidence. The timber roof over this is of a great age, and two bays have survived the recent fire. This is illustrated in Fig. 68, and is of seven cants assembled by open notched-lap-joints. It has been previously published (C. A. Hewett, 1971, 84–88; and 1974, 11), and needs little more than illustration for present purposes. The crown posts were later intrusions, fitted at such time as the canted roof had become insecure, and they were considered to date from the late 13th century by Mr. H. Forrester. It may be seen on the site that some tie beams' ends were left circular where embedded in the rubble masonry of the walls, and were squared only where they emerged into view. What appears to be a strictly analogous tie beam's end may be seen above the vaulting in the nave at Winchester Cathedral, where it was sawn off flush and left *in situ* when the nave was re-roofed at an early date.

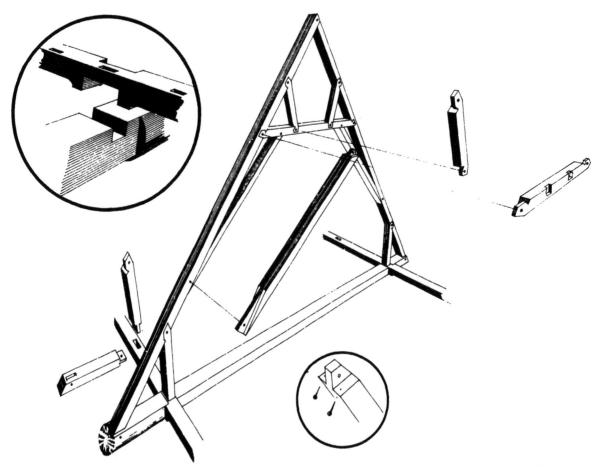

Fig. 68. Roydon parish church, nave roof, use of spikes inset. Inset, top; circularity of embedded tie beams.

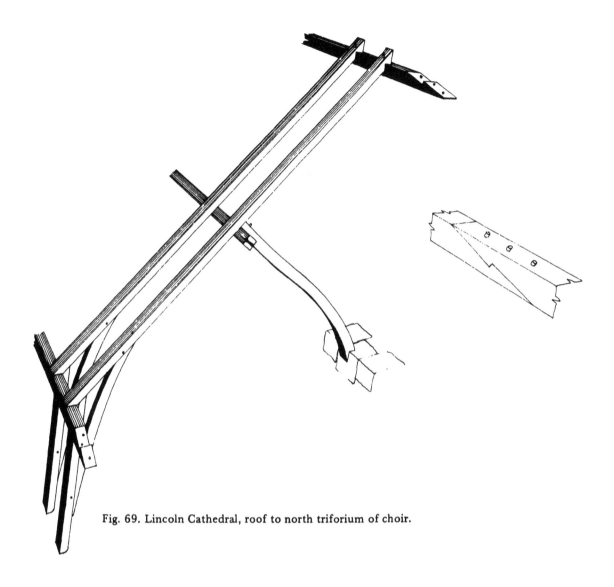

Fig. 69. Lincoln Cathedral, roof to north triforium of choir.

The Roof of the North Choir Triforium, Lincoln Cathedral

This is a very graceful design, and one relying absolutely upon the stability of the masonry. It is illustrated in part in Fig. 69. The choir and east transept were completed between 1192 and 1200, under Geoffrey de Noyers and Richard Mason (J. H. Harvey, 1961, 141), but this roof may not be original to this part. The wall- and top-plates were scarfed with the ancient *trait-de-Jupiter*, as shown; and the sinuous strut bird's-mouthed over the masonry, together with the use of curved bracing and wall pieces at the eaves, were anti-sagging devices.

A disastrous fire in 1174 must have destroyed the original roof (necessarily of unknown form) to this tower, and the Caen stones within it show the pink discolouration associated with that intense heat. The rebuilding of the east end was begun in 1175 under William of Sens, and William the Englishman continued with it from 1179 until its completion in 1180 (J. H. Harvey, 1961, 121). In view of the fact that Canterbury Cathedral is remarkably well documented, and that no building activities of importance were recorded between 1184 and 1220, it has been suggested that the surmounting spirelet (which is small even by parish church standards) may be of the 12th century (C. A. Hewett and T. Tatton-Brown, 1977, 129). The framing is illustrated as Fig. 70, with an example of its 'secret' notched-lap-joints shown separately to the left of the drawing.

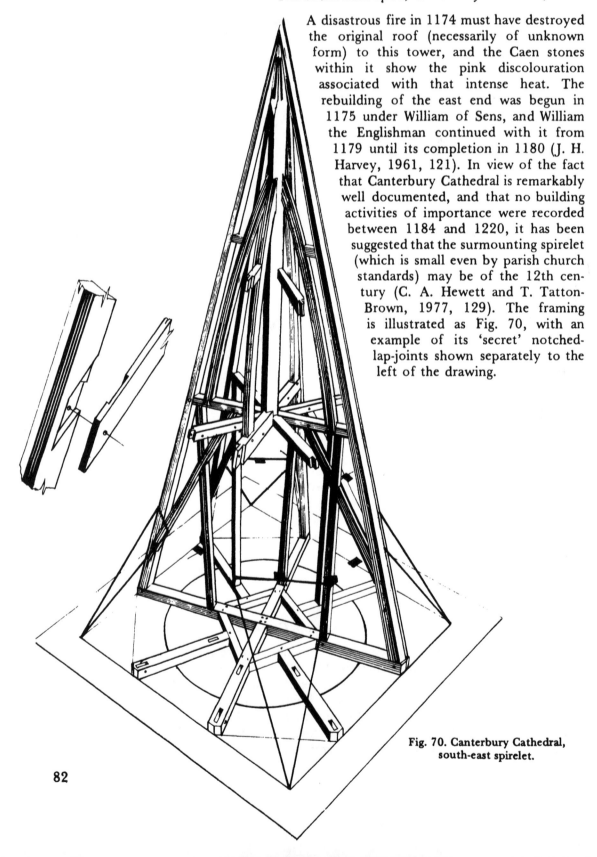

Fig. 70. Canterbury Cathedral, south-east spirelet.

82

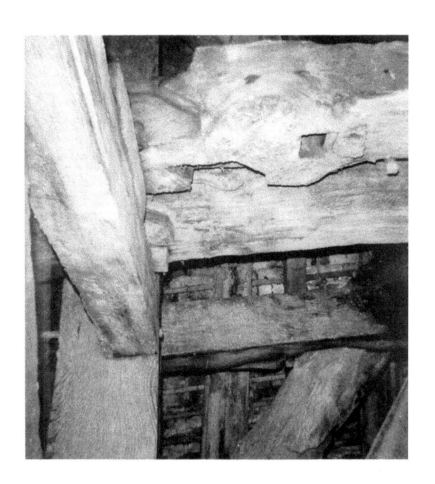

I Matrix of a notched-lap-joint on a re-used
timber in the belfry of the parish church at
West Bergholt, Essex.
(Author's photograph.)

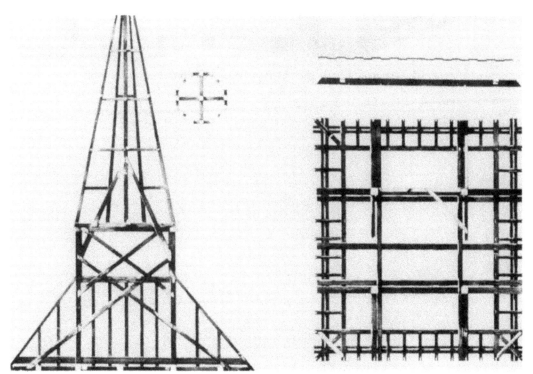

II Plan and vertical section, scale-drawn, of the spire of the parish church at Upminster, Essex.
(Author's watercolour.)

III   The roof of Little Hall, Merton College, Oxford after restoration.

IV A queen post in the roof of Little Hall, Merton College, Oxford. (Plates 3 and 4 by J.W. Thomas, ARPS, Oxford.)

V Vertical section of Salisbury Cathedral Spire. (Photograph by John McCann.)

VI *(above)* The roof of Westminster Hall.
(National Monuments Record.)

VII *(right)* The roof of Westminster Hall showing the
longitudinal arcades. (National Monuments Record.)

VIII The Chancel roof of the parish church at Saffron Walden, Essex. (Photograph by Frank Joel).

IX   Blackmore Belfry. (Photograph of author's watercolour)

X   The loading arrangements for joist-shear tests. (Photograph supplied by Dr. D. Brohn, Bristol Polytechnic.)

XI *(above)* The barn, Prior's Hall, Widdington, Essex.

XII *(right)* Joint from the church of St. Thomas the
Apostle, Navestock, Essex.

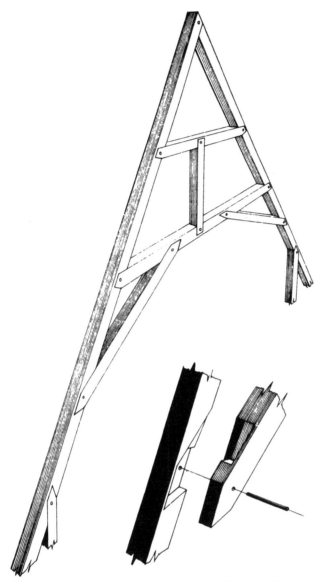

Fig. 71. Wells Cathedral, nave high-roof after *c*.1213; secret
notched-lap, exploded, inset to foreground.

The High-Roof over the Nave, Wells Cathedral, Somerset

This roof was continued after the Interdict, and continued westward to its
termination at the west front during the life of Adam Lock, the architect. The
transverse design was unaltered, and the rafter couples immediately west of the
point where building had ceased in 1209 are datable to 1213 and later. The
difference between the two lengths of roof was only that of joint type, the 'secret'
form of the notched-lap-joint being used after the break. It is possible that Lock or
his carpenter had invented it during the years of the break, when little but site
maintenance could have occurred. An example of a secretly jointed roof couple
with the joint enlarged and exploded forms Fig. 71.

Fig. 72. Peterborough Cathedral, west door leaves.

The West Doors, Peterborough Cathedral, Cambridgeshire

These are illustrated in Fig. 72, and are the central pair in relation to the west front, which is datable to between 1230 and 1293 (J. H. Harvey, 1961, 149). The use of diagonally-placed ledges may be an allusion to the dedication of the great church. Several decorative features, such as the dog-tooth ornament on the ledges and the naturalistic foliage carved on the capitals of the two heads, indicate a date near the end of the given date range.

The Church of St. Laurence, Upminster, Essex

Of this the Royal Commission (R.C.H.M., 1921, 160) stated that the west tower was of *c.* 1200 and the bell chamber probably of a slightly later date. The carpentry of the surmounting spire is illustrated in Plate II, which shows both plan and vertical section to scale. The secret notched-lap-joint was used to secure all the diagonal bracing, which forms saltire crossings—for which long slender timber was not available and two lengths were used, as at Cressing wheat barn. Another datable feature which occurs on the wall plates of the elaborate platform assembly on which the whole was mounted is the hewn fillet, over which the sole pieces were trenched, as has been noted at Lincoln.

The Church of St. Mary Magdalene and St. Mary the Virgin, Wethersfield, Essex

The west tower of this church has been described as massive, and its spire as of more German than English form (Sir N. Pevsner, 1956, 386); it is ascribed to the early 13th century by the cited authority and to the end of the 12th century by the Royal Commission (R.C.H.M., 1921, Vol. I, 333). It is without buttresses and incorporates several interesting pieces of carpentry, among which is the first floor, mounted as it is on a samson post; but foremost in the present context is the spire itself. This, like that at Upminster, is set on a platform structure, and has two stages of diminution. The framing is shown as Fig. 73 in which the near corner has been omitted. There are in this work four secret notched-lap-joists, indicating a true date after 1213.

Songers, Cage Lane, Boxted, Essex

At the time of writing this is the most remarkable small timber-framed residential building generally known; examination proves it to have been built as a two-bay open hall with one storeyed end, these features being designed within a diminutive ground plan measuring only 26ft. by 21ft. Today the structure is five bays long, having had two further storeyed bays added to its eastern end, probably during the 14th century, at which time it seems that the open hall was extended from two bays to three. The hall was laterally divided by a first floor during the mid- to later 16th century, when the usual red brick chimney stack of that period was also intruded.

The original framing is illustrated in Fig. 74, in which all later accretions have been omitted, as has the framing of the side walls, for the sake of clarity. The roof is heavily sooted, and is fitted with collars only; it is thatched now, and there is no evidence for any alternative cladding in earlier times. There is absolute structural evidence for the rafters of the hipped ends having been radially disposed, like the blades of a fan—a feature noted in respect of the Farmhouse, Thorpeacre, Loughborough, Leicestershire (T. H. Rickman, 1968, 150-3).

The first build was assembled with open notched-laps, which were used in an astonishing profusion; as a count from the illustration confirms, these number thirty-six. The groundsills, which cannot be shown to be replacements, are trench-jointed at right angles, and superimposed upon one another in a manner

Fig. 73. Wethersfield church, spire framing.

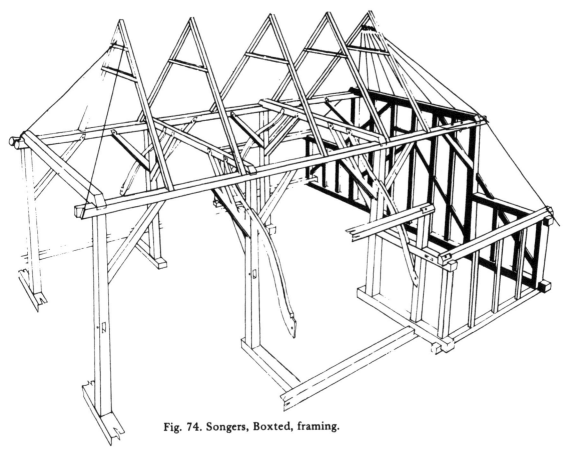

Fig. 74. Songers, Boxted, framing.

that conforms to a decline of the site level at the western end. One top plate is the entire length of a small oak trunk, and terminates at its north-eastern end in the fork of two major crown branches, as shown in the drawing. The opposite top plate is scarfed at its south-eastern end by means of a stop-splayed joint with two large face pegs and under-squinted abutments. At the western end, which was moved to its present position at the time when the two bays were added, an internally visible windowsill was either used at the outset or was intruded; in either event it is now serving the latter capacity, and is of interest because it shows a now disused matrix for a squinted lap joint that was pegged but not notched, resembling those visible and in use in the Rhenish helm of Sompting, Sussex. As illustrated, the ground plan was T-shape, a feature not noted elsewhere.

The Barn, Siddington, Gloucestershire

This barn forms the south-western boundary of the churchyard at Siddington, near Cirencester. Both the Siddingtons, Upper and Lower, were always in lay hands, but the church passed to the Knights Hospitallers and was valued at 20 marks per annum in the 14th century (L. Larking, 1957, 28), two-thirds of the conventional value of a knight's fee. This, as Rudder seems to imply (S. Rudder, 1779, 661) may mean a small 'rectorial manor', or the barn might be a true tithe barn, indicating a high capital yield. Rudder ascribes the donation of the church to one Jordan de Clinton who, according to H. Fynes-Clinton (*Memorials of the Clinton Family,* no date, table at end), was the younger son of a younger son, whose elder cousins (none of them connected with lay manors) flourished in the 1160s. The gift, especially if by testament, would seem to have been made *c.* 1200.

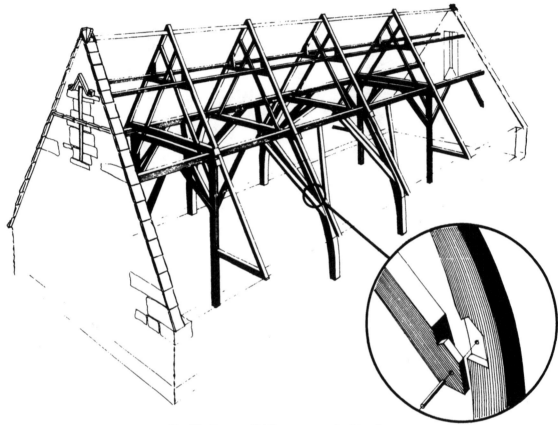

Fig. 75. Barn at Siddington, notched-lap inset.

The building is stone-walled and designed in five bays, each measuring about 18ft. by 36ft., and having four transverse timber frames to support its roof. The masonry of the eastern gable is apparently undisturbed and the top plates are embedded in it; their ends are visible on the outer face, as are those of the timber corbels provided for the mounting of the terminal arch braces. The existing midstrey is a later, though ancient, addition. The two end frames are aisled and posted, and equipped with double tie beams which clasp the top plates; scissored braces which pass the ties low enough to meet the posts; collars with side struts to trap the side purlins, which are mounted in the flat position, and a ridge piece cunningly trapped between the rebated faces of the principal rafters at their apexes. These frames were assembled by means of open notched laps. The central pair of frames are base-crucked and have a reversed-assembly at their tie beams, with the same complex frame design above that point. The braces from the base crucks to the lower tie beams also have open notched laps at both ends, as detailed in the inset of the drawing (Fig. 75). The top plates are scarfed directly over the posts' heads or those of the base crucks, having a through-splayed and tabled joint and face pegs.

88

Roof to the Vestibule, the Chapter House, Lincoln Cathedral

Among the smaller roofs of Lincoln the one illustrated in Fig. 76 is of interest. It was assembled by both tenoning and notched-lap-jointing, the latter being of the 'archaic' open type. Tie beams were fitted to every fourth couple, and iron spikes were used to affix the tops of the ashlar pieces. The provision of tie beams to a roof without wall plates may seem illogical, but reflection upon the matter makes it clear that their purpose was to contain the spreading moment at short intervals. The work of building this vestibule occupied the years 1220–35, and the master responsible is uncertain.

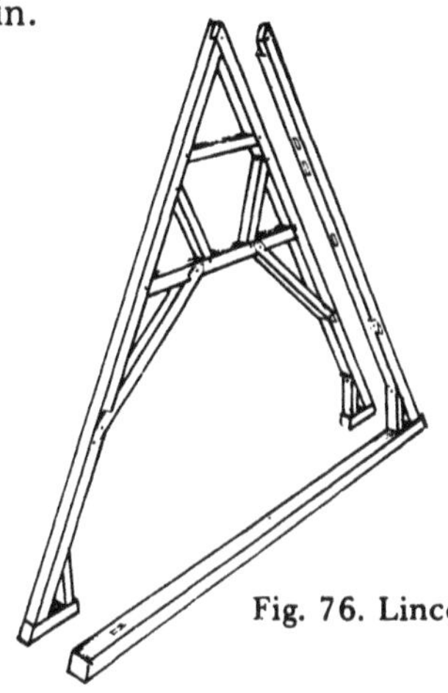

Fig. 76. Lincoln Cathedral, roof of chapter house vestibule.

Romsey Abbey, Hampshire

Some distance south-west from the abbey church, in what is now a private house, survives the timber roof of what is thought may have been the refectory. This is a design of seven cants, probably dating to *c.* 1230, and assembled by notched-lap-joints. The soulaces have 'open' notches at both of their ends, but the collars have 'secret' notches (Fig. 77). This is remarkable in that the strongest, but most expensive, form was used very judiciously and then only at the point judged to be most severely stressed. This is, at the time of writing, the only known example of an eclectic use of notched-lap-joints.

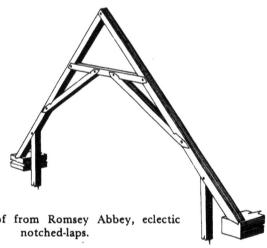

Fig. 77. Roof from Romsey Abbey, eclectic notched-laps.

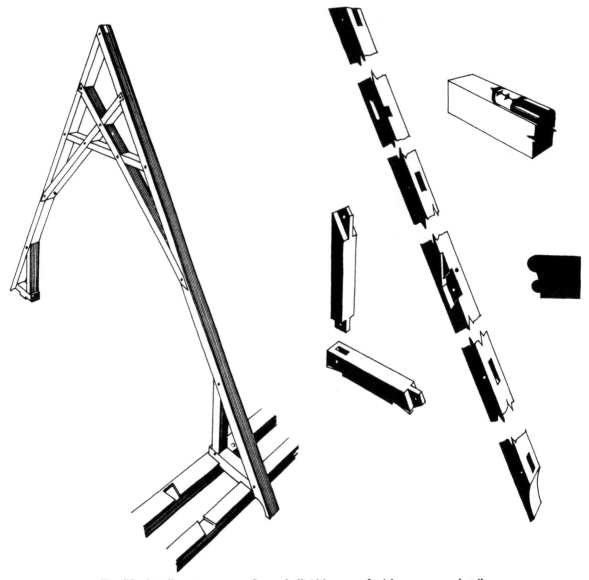

Fig. 78. *Capella-extra-muros*, Coggeshall Abbey, roof with numerous details.

The Capella-extra-Muros, Coggeshall Abbey, Essex

This chapel has survived long use as an agricultural building and is in a remarkably complete state; it was built as the chapel outside the walls of the immediately adjacent abbey of Coggeshall. It also survived a 19th-century restoration without losing its original roof, comprising 25 couples, scissor-braced and collared. It was built during the term of Abbot Benedict from 1219–20. The structural principles of the roof are illustrated in Fig. 78, in which the joints spaced along each rafter are shown separated to the right. The bases of the scissors were secured with the unwithdrawable secret notched-lap-joint, which had possibly been transmitted from Somerset within the lapse of only five years.

90

The North Choir Transept, Salisbury Cathedral, Wiltshire

The high-roof of this transept at Salisbury is without doubt the original carpentry, the three eastern arms there being completed, as Dr. Harvey has shown, by 1237. In the opinion of the writer this is one of the finest timber roofs in the high-roof category—which were never intended to be seen—among all those surviving in English cathedrals. It sustains an interesting comparison with that over the nave at Lincoln, in that it has the slender pairs of chamfered posts placed beneath its lower collars, but beyond this point similarity ends.

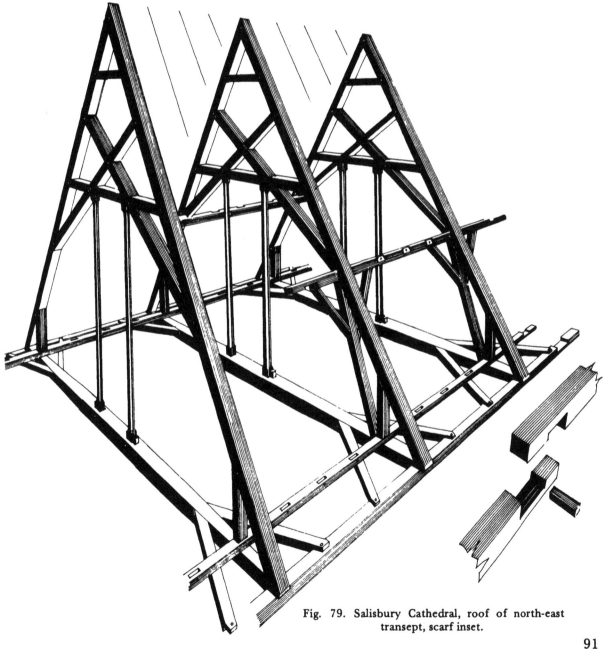

Fig. 79. Salisbury Cathedral, roof of north-east
transept, scarf inset.

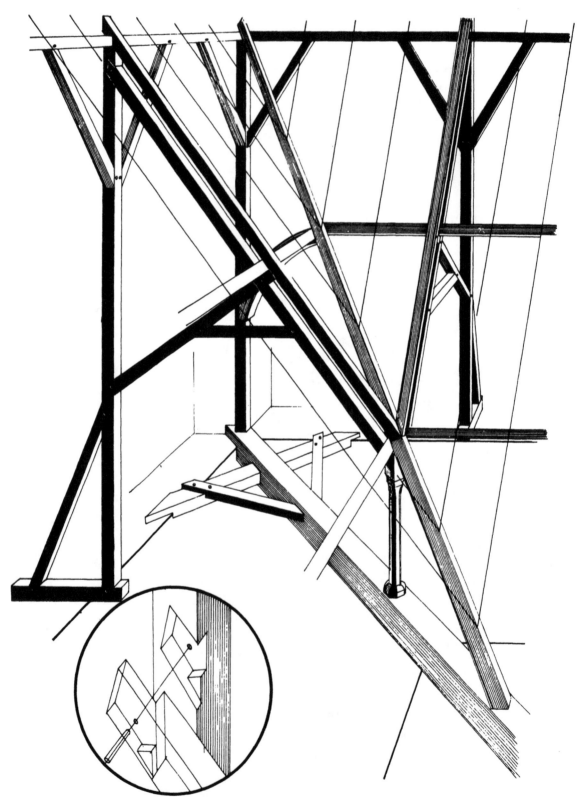

Fig. 80. Salisbury Cathedral, return of triforium roof.

The tie beams were placed between every three common couples, and were correctly tapered at their ends so as to provide maximum strength where stress was greatest; in addition they were branched into three arms at each end to provide a width tie for every couple. Secret notched-lap-joints were used for their ends. Side purlins were fitted, mounted on anti-racking canted posts and arris-trenched to receive the purlins; and herein, seemingly for the first time, compressive scissor braces were used, each comprising five short timbers chase-tenoned together. Other features may be seen in the drawing. The scarf used to combine the lengths of wall plate is unusual and is shown at the bottom right, Fig. 79. It has been found in only one other English context—the wall plates to the nave roof at Winchester Cathedral, not far removed in either space or time. This fact implies the common involvement of an architect, or master carpenter, whose name has eluded the records.

The Roof of the North Triforium, Salisbury Cathedral, Wiltshire

The most impressive lean-to roof surviving in any English cathedral, this is a most remarkable essay in carpentry attributable to either Elias of Dereham or Nicholas of Ely, and datable to between 1225 and 1237 (J. H. Harvey, 1961, 157). A deliberate approach to the building of an asymmetrical ridged roof was made here, resulting in a design that could stand alone and quite independently from its triforium situation. Being so designed it has anti-racking timbers incorporated in the framing, and these took the form of triangulating braces at the posts' tops, as illustrated. The turning of the triforium returns was even more spectacular, and an example of this is given as Fig. 80. This is drawn as though viewed from the outside, and the hatched timber ascending from lower right to top centre is the valley rafter, mounted upon a basically logical yet highly ingenious transom that bisects the return. Some secret notched-lap-joints were used in this roof, at the points where the under-rafters meet the posts, and one of these is shown in the encircled inset diagram.

Door Leaf at Waltham Abbey, Essex

The head of a right-handed door leaf is preserved here as a museum artefact, both faces of which are illustrated in Fig. 81. The harr durn was scarfed, literally, as distinct from jointed; and forms of notched-lap-joints were used for assembling its portcullis-type rear leading. The 'shuts' were worked into a sally or tace, with a concomitant rebate upon the missing leaf—a rare example of experimentation in this field of carpenters' activities. The cusping worked on the frontal muntins is so asymmetrical as to question the period to which it should be ascribed; but the arcature of the durn indicates that the style was pointed and Early English is suggested, since this leaf probably belonged to the external west portal which was earlier than the early 14th-century lady chapel (Sir N. Pevsner, 1956, 371).

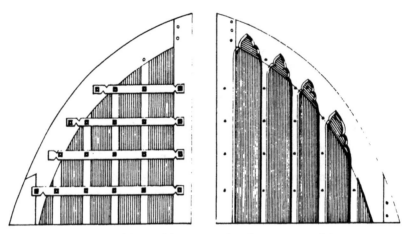

Fig. 81. Waltham Abbey, west door leaves, as surviving.

The North Door Leaf, Dore Abbey, Hereford

This door is built of v-edged planks of oak with three chamfered ledges across its rear face and a trefoiled head made of two sections; the whole is peg-fastened with circular pegs of large diameter. The masonry reveals of the aperture have one prominent mould, the roll with frontal fillet, dating from between *c*. 1219 and 1250 (H. Forrester, 1972, 31). This is illustrated in Fig. 82, in which the decorative ironwork is necessarily emphasised, although it is not the proper subject of this work. An account of a comparable door, insofar as framing is concerned, is published in respect of Belchamp Walter in Essex (C. A. Hewett, 1974, 107).

Fig. 82. Dore Abbey, door leaf.

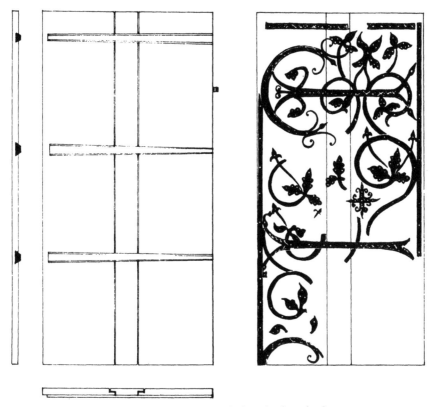

Fig. 83. Eastwood church, door leaf.

The South Door, Church of St. Laurence and All Saints, Eastwood, Essex

Few positive developments occurred in the carpentered construction of door leaves, but one highly-skilled and rather delicately-framed method was incepted, perhaps at Durham Cathedral, for the access doors to the cloisters; it was also used at Eastwood church, where the example is of even higher quality (Fig. 83). This was made from three oak planks, two wide and one narrow, with rebated edges; four ledges of a dovetailed cross section were provided, all tapering in their widths, and were driven into suitable trenches on the rear face. The main strength of this door, like that of the Durham example, is provided by the lavish application of decorative ironwork. One of the straps bears a Lombardic inscription: *Pax regat intrantes eadem regar egredientes*—'may peace rule those entering and also those leaving'. The date is uncertain, but the earlier years of the 13th century are proposed. The wrought iron is particularly important because the leaf shapes wrought are oak leaves—a reference to local flora, as in the *Leaves of Southwell* (Sir N. Pevsner, 1945, 1-71).

The South Door Leaf, Church of All Saints, High Roding, Essex

The style of the fabric, a nave with chancel, is 13th century, to which period the Royal Commission (R.C.H.M., 1921, Vol. II, 133) ascribed the door. This is deceptively simple at first sight, but proves upon examination to be one of the most soundly and thoughtfully constructed door leaves known. Both front and rear aspects are shown in Fig. 84, together with an exploded structural drawing. The iron strap hinges retain the main characteristics of those from the preceding century, including divided curls and the anchor shape, supporting the date ascription. The method of assembly is, again, unique to date; as shown in the cross

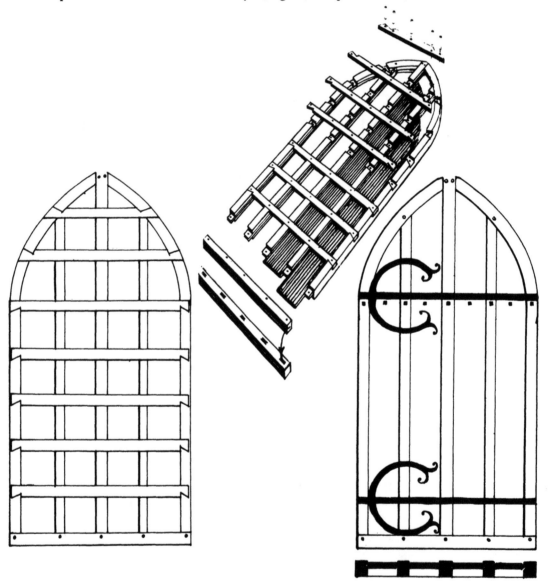

Fig. 84. High Roding church, the door leaf.

section the stiles and durns have v-grooved edges and the ledges halved across them are of lesser thickness, so that the whole frame could be assembled before the planks were slid into place. Finally the bottom rail was fitted on with stub tenons and pegged. This is truly a carpenter's door, and its ironwork provides nothing more than the hinges and some decorative contrast of materials.

The Church of St. Martin, White Roding, Essex

The Norman nave of this church is roofed by the important assembly illustrated as Fig. 85. The date of this work is not known, and it can only be calculated by comparison with dated examples. The carpentry cannot be that which was originally built for the fabric, yet it must be a replacement that had become necessary during the 13th century by virtue of the physical inadequacies inherent in its predecessor. A date between 1250 and 1300 is proposed in view of the form of scarf joint used, which is encircled in the drawing and discussed further under the heading of jointing (pp. 263-271).

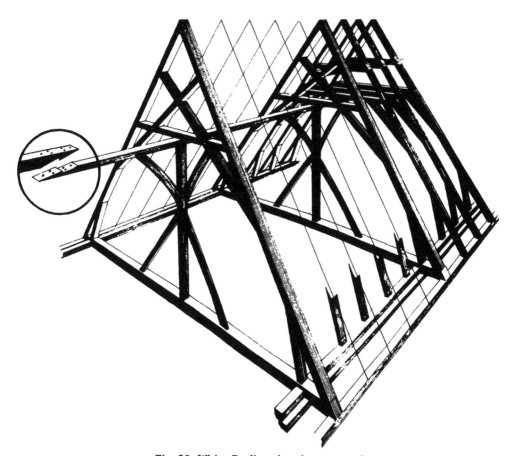

Fig. 85. White Roding church, nave roof.

Paired wall plates were fitted in this example and the internal plate mounted an ashlars' plate. The three couples illustrated in the Figure that have double collars instead of scissor braces were repairs, fitted after some ancient damage to the nave. They help to date double collars as a 14th-century phenomenon, insofar as parish churches were concerned.

The Church of St. Mary, Aythorpe Roding, Essex

The western bell turret here is unusual in that it was mounted upon two transoms which span the nave and are braced downward, probably to wall pieces, although thick plaster prevents proof of this probability. The entire frame is illustrated in Fig. 86, which at upper right shows the type of timber outsets used to mount both top plates and tie beams. This so closely resembles the early upstands used elsewhere (e.g., Cressing wheat barn) before the inception of the jowl that a date ascription of *c.* 1260 is tenable. It is considered that the bell turret must date from the initial completion of the church, and the framing method used; the use of prick posts centrally in each wall with saltire braces trenched through them obviously derives from the pattern set at Bradwell-juxta-Coggeshall during the 12th century. The use of subtly curved timber here denotes the Early English style. The long straight-edged increase in the posts' heads clearly presages the impending jowl, and may be seen to derive from the short but similar form at Bradwell. The humped inner or spandrel edges of the arch braces are interesting here, since they establish this somewhat rare treatment of braces as late 13th century.

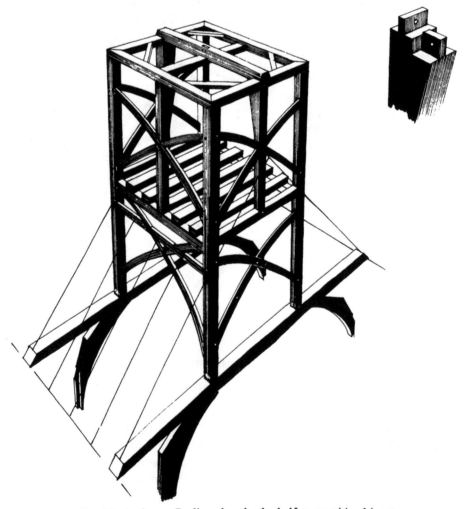

Fig. 86. Aythorpe Roding church, the belfry, quasi-jowl inset.

The Early English Period—Summary

The works described belong to the latter half of this period, insofar as it is possible to assess their dates. The Great Transition from Norman into Early English had been effected, and these works represent the established style—or, more accurately, they should reflect the influence of the new style upon carpenters of its time. The cross-sectional designs of cathedral roofs, which had now become high-roofs, since they were always above vaults, are represented by the nave roof of Wells, which was a direct development from the earlier roof of Waltham Abbey: but there is evidence for more than one course of development. The series of roof designs for Lincoln Cathedral was progressing and two almost unrelated designs were produced there for the transepts of the eastern crossing.

The notched-lap-joint was consistently used, however, over an area as wide as is covered by the examples. It appears to have been perfected at Wells during the years of the Interdict, but as we can see from the examples its best form was not adopted elsewhere. The secret form of this joint was used at Canterbury for the south-east spirelet and at Westminster for the eastern arms of the Abbey Church, but never at Lincoln Cathedral; however, quite minor buildings in eastern England such as the two spires at Upminster and Wethersfield in Essex took advantage of the improvement almost immediately. The only evidence for purely personal judgements as to what was better, and how developments could best be combined, is provided by the roof of the building at Romsey Abbey in Hampshire, where both open and secret laps were employed. This is a matter of great interest because the secret form was obviously more costly to produce, and for that reason a wide advocation of both types of joint was to be expected, but surviving buildings indicate an almost unquestioning acceptance of the new form. Lincoln Cathedral is the exception that proves the general rule, and is difficult to explain.

With the acceptance of secret notched laps by those carpenters who employed them, came the visual simulation of continuous long lacing timbers by closely jointing two short ones, as noted at Upminster spire and the Cressing wheat barn. It is important that these two uses of this device were in very different situations—the expedient (if that is what it was) was visible in the barn, but high in the spire it was unlikely that anyone would ever see it; therefore it may be that single long timbers were less readily available during this period, or that their cost was occasionally thought prohibitive. The use of wall plates having hewn face fillets for the location of the sole pieces seems to belong to this period, and seems also to have enjoyed a widespread advocation that was thinly scattered rather than universal; this again must have been controlled by costs. The principle involved was that noted of the Waltham Abbey roof—cross-cogged tie beams, refined and improved, at great expense, and applied to sole pieces. A curious fact of the roofing in these years is the invariable recognition of the value of side purlins for lean-to roofs, themselves a type of roof unable to rack. During the same period purlins were fitted into the pitch planes of many roofs that have survived, such as that of the Siddington barn; but it seems evident that they were thought necessary in lean-to roofs merely to retain their 'flatness' of plane, and were only sporadically fitted into ridged roofs for the different reason of inhibiting racking movements.

A structural principle that was retained throughout was that of the passing brace. It seems apparent that this had originated at least as early as the 11th century, and in the earliest examples such timbers were of great lengths, as formerly in the Cressing barley barn; but during these years they were contrived from shorter timbers, jointed usually on the main posts' flanks. It may be, in the light of this fact, and that of the Lincoln Cathedral rafters being scarfed together from two timbers, that the intensity of building operations at this time strained the available resources of long oaks. *It is surprising in this context that passing braces were not scarfed together, the splays of the joint being housed in the post's flank trench —but no example of this has as yet come to light.*

The examples furnish clear evidence for both the continuation of traditions, as in the high-roofing of great churches, and for the pursuit by individual craftsmen of original thought and design. This statement at once seems trite, but it is justified, and after the ensuing Decorated period has passed it is no longer true. Carpenters' problems did not change much, if indeed they changed at all, and their material was fixed; as a result the number of efficient methods for framing the few required items was soon established. The high-roof of Salisbury's eastern north transept is an astonishing example of developing and combining structural principles which seem to have been derived from a synoptic view of the best of carpenters' works until that date from all parts of the country. This is a proposition that has always been contested, or queried, because it is generally held that communications in 13th-century England were inadequate, and that news of inventions could not have spread rapidly; but the evidence implies that this was not the case, and evidence must be assessed as such. With the triforium roofs of Salisbury it seems that entirely novel concepts were introduced.

It may well be that earlier lean-to roofs, long since gone, led up to this design; but it is suggested that, as exemplified by the High Roding door and the former triforium roof at Norwich Cathedral, a few medieval craftsmen combined ingenuity with a genius for sheer invention to a degree that produced designs of which the majority were incapable. It cannot be established now, when only the small number of carpenters' works that have survived can be assessed, whether or not the designer of the High Roding parish church door was the innovator of its type—but no known cathedral door is similar. An established tradition, begun in the early 11th century at Hadstock, had been planked door leaves strengthened by a rear frame of ledges; but at High Roding there appears a new, superior concept— the framed door leaf into which the planks were inserted, the resultant strength deriving almost equally from the frame and its infill. Cathedral examples that may prove similar when examined in depth, such as the cloister doors at Hereford and Gloucester, are of much later dates.

Again, it is true of the examples selected that concessions to the visual style, with its lancet arches, are to be found in a minority of buildings; however, this aspect of the matter is less important than it may seem. Wherever a decorative treatment such as a moulding, capital or base, or an arch that was intended as a displayed architectural feature had a legitimate place in the design, it conformed strictly to the obtaining style. An apparent exception to this is the wheat barn at Cressing, wherein curved timbers were not used in any principal capacities; but this is due to the same adherence to purely structural principles that had excluded arcuation from cathedral high-roofs like Wells, Salisbury and Lincoln, themselves designed to cover substructures whose basis was the arch.

The explanation of this dissimilarity between carpenters' and masons' designs lies naturally and obviously in the nature of their materials. Carpenters must think in terms of unit length of timber, units well suited to resist every building stress known—while masons were reduced to terms of stone, quarried and worked in the most practicable sizes and forms, and basically capable of resisting only compression. When bracing an arcade the carpenter did not normally envisage a solid spandrel because his timber did not require it for stability, and it was only by a long-drawn-out process that carpenters in the south-eastern area came to accept the principle of curved timber, which, once adopted, was to exclude further use of straight timber.

One of the most important of carpentry developments was beginning during these years—the use of jointing complexes, of which the upstand used in the wheat barn is an early indicator. Previously it seems that any convergence of three or more structural timbers, such as posts, top plates and tie beams, had been seen as a succession whereby the second was jointed to the first, and the third to the second item. With upstands this changed radically, and each of the three timbers was equally jointed to the other two. From this basis the jowl was derived and became absolutely invariable, but it is not yet established how long its development took— we only know that it was perfected by the end of the 13th century.

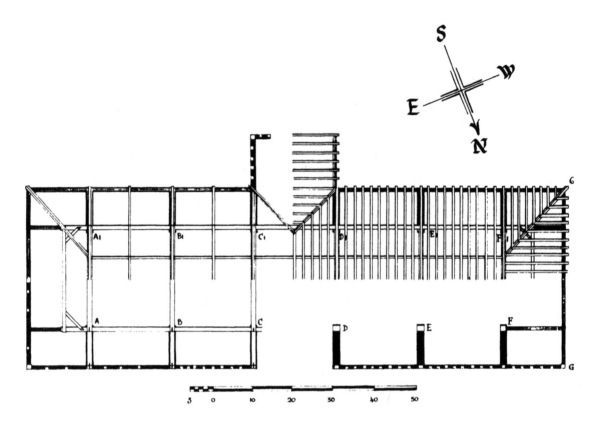

Fig. 87.   Scale-drawn plan of Cressing wheat barn.

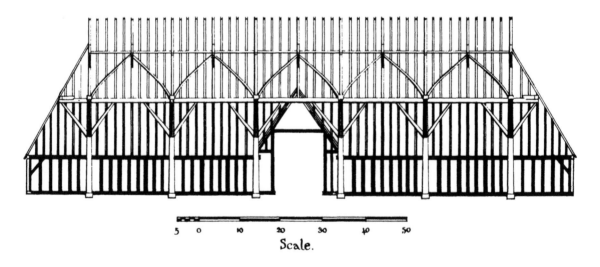

Scale.

Fig. 88.   Scale-drawn longitudinal section of Cressing wheat barn.

102

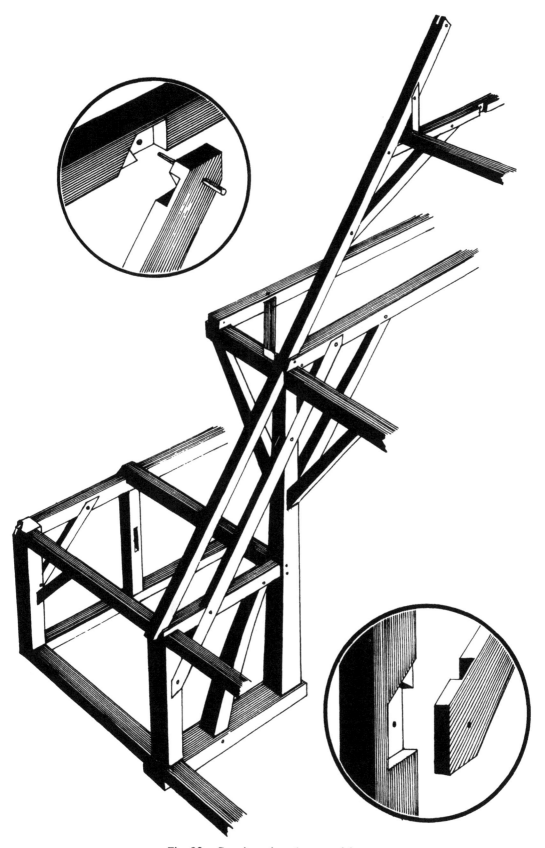

Fig. 89. Cressing wheat barn, end frame.

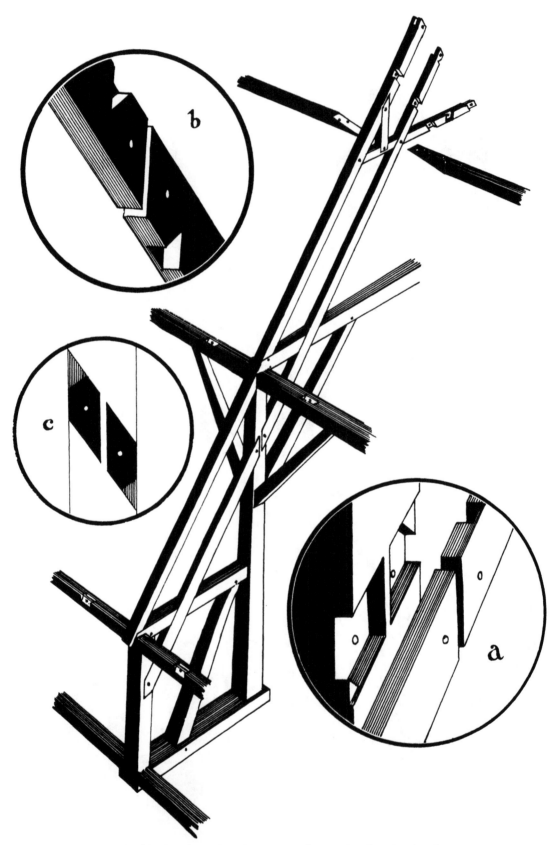

Fig. 90. Cressing wheat barn, cross frame, showing simulated
long timbers.

Chapter Five

Examples from the Decorated Period (*c.* 1250 to *c.* 1350)

The Wheat Barn, Cressing Temple, Essex

The younger and better preserved of the two great barns on this site has been carbon-dated to *c.* 1255, as was previously published (C. A. Hewett, 1967, 48–70; and 1969, 59); only new information need be added to the earlier statements, with the same illustrations suitably amended. Since this barn is more comprehensible than its older companion the scale-drawn plan, with its stud work corrected to original frequency, is again used as Fig. 87, and the accompanying longitudinal section as Fig. 88. In Fig. 89, diagrammatic perspectives are given for an end frame; the timbering of the original eastern wall has now been added to the drawing. It is important with regard to this, and the date, that the top plate at eaves level of this original end is made from a single tree which is among the longest single oak timbers known in Essex at this time. An intermediate frame is shown as Fig. 90; the use of separate short lengths to make apparently single and continuous passing braces is analogous to that in the Upminster spire, and is illustrated in the various circular insets to the drawing. The upstands used to unite the posts to the top plates are important since they are the first tolerably firmly dated examples of the kind. It is from these that the more elaborate jowl was developed, and their presence may date any building to the mid 13th century.

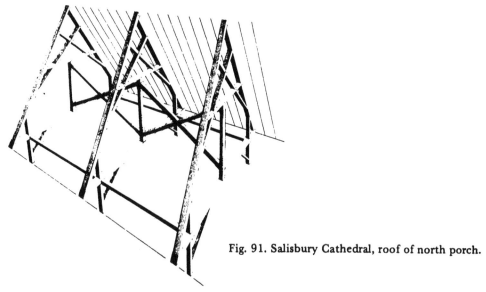

Fig. 91. Salisbury Cathedral, roof of north porch.

The High-Roof of the North Porch, Salisbury Cathedral, Wiltshire

All the roofs at Salisbury were probably completed before the west front, i.e., before 1266, and the building of the nave westward from the crossing occupied the years between 1237 and 1258; the porch, therefore, may have been roofed *c.* 1250 (J. H. Harvey, 1961, 157). However, a remarkable introduction was made in the form of crown posts, without collar purlins, which were saltire-braced—a device designed to contain the racking of the couples within any one bay in which it might commence. Additionally, this roof had compressive scissor braces on every couple, the type apparently incepted when the eastern arms of Salisbury had been roofed. Strangely, this development seems to have had little effect, and no comparable or derived crown-post roofs are known at the time of writing (Fig. 91).

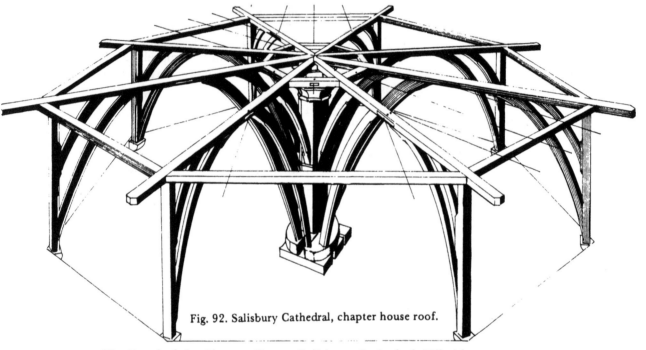

Fig. 92. Salisbury Cathedral, chapter house roof.

The Roof of the Chapter House, Salisbury Cathedral, Wiltshire

The chapter house was probably complete by 1266, although positive evidence as to its completion date or designer are lacking at the time of writing. This roof is illustrated in Fig. 92, without indications of the stone sub-structure. The design has a central pier to support its vaulting, and this is continued in timber for the support of the roof, which is almost flat, but subtly conoid. The profile of the carved capital applied to this post is shown in Fig. 93.

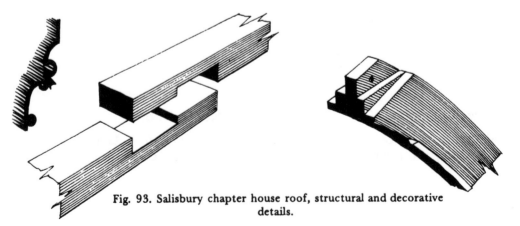

Fig. 93. Salisbury chapter house roof, structural and decorative details.

The total weight is divided between the pier and the returns of the octagonal
·stone carcase, in each of which a post is mounted upon a corbel; the rafters are
radially disposed and are each supported by no less than four well-curved braces,
two from the pier and two from the posts. Purlins were tenoned into the rafters'
edges. The scarf joint used for the top plates of this work is also illustrated in
Fig. 93, together with the type of spurred tenons used for the great number
of braces.

Chichester Cathedral, Sussex

The high-roof here is a remarkable one making intelligent use of the branching
ends of both raking struts and crown posts, but not applying the same principle
to the tie beams, which have eight rafter couples between them. Both crown-post
and raking-strut ends are cut into pronounced jowls having straight, oblique outsets,
resembling the earliest types derived from the upstand, such as those at Priory
Place (Fig. 115). Two great fires are recorded at Chichester, the latter having
occurred in October 1187, after which the great church was re-roofed and the stone
vaults added; reconsecration took place in 1199. The collapse of the central tower
and spire did little damage and the repairs executed at that time (1862) are easily
detected. The actual date of this roof can only be deduced. It must obviously be
later than 1199 because it exhibits jowls, which could not have been used before
plain upstands; and since it employs the branching timbers in compression that
were first used in extension at Salisbury and Westminster it should, on the basis of
typological succession, be placed later (Fig. 94).

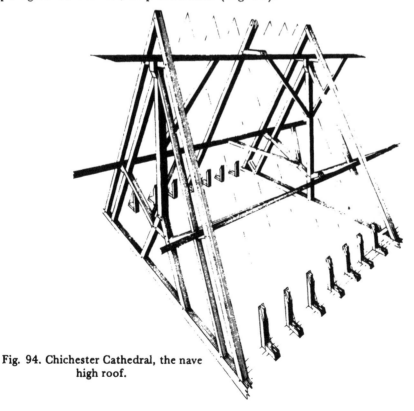

Fig. 94. Chichester Cathedral, the nave
      high roof.

The Bury, Clavering, Essex

Of this site Morant says: 'It was originally in the King's disposal; and went, commonly, along with the Lordship of Clavering, which was holden of the King, *in capite*'. The surviving manor-house here is so important, and so dissimilar to others of the 13th century, such as Place House at Ware, that a detailed account of its occupants is justifiable. For this research I am indebted to Mr. R. G. E. Wood, Historical Advisor to the Essex County Council, whose findings are quoted for what seem to be the relevant years. 'From before 1086 to 1289 there is nothing to suggest that any of the lords of Clavering resided on the manor. All had larger estates in other parts of England: Suene and Robert of Essex at Rayleigh, and Alice's second husband Roger, as well as their son and grandson, in Northumberland where they were busily adding to the defences of their castle at Warkworth, which Henry II granted to Roger in 1158. The brief Inquisition after the death of Roger fitzJohn in 1249 makes no mention of a manor house at Clavering. The only exception was Alice of Essex, who may have lived at Clavering during the period between the death of her first husband (1141 or earlier) and her marriage to Roger (possibly after 1158).

'For sixteen years Roger fitzJohn's son Robert was a ward of Henry III, who granted custody of Robert's lands (including Clavering) to his own half-brother, William of Valence. The king spent the night of 11–12 March 1251 at Clavering, and

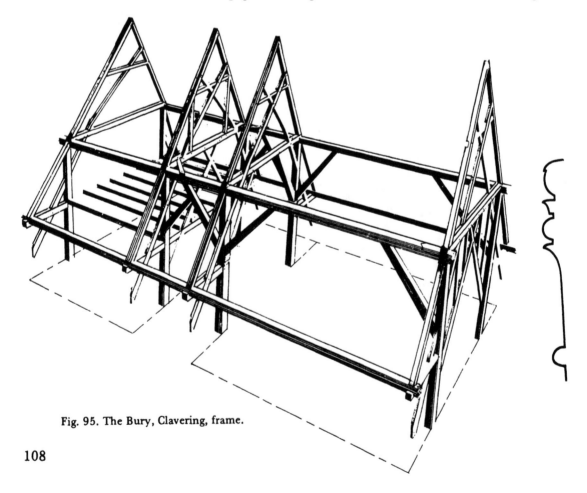

Fig. 95. The Bury, Clavering, frame.

108

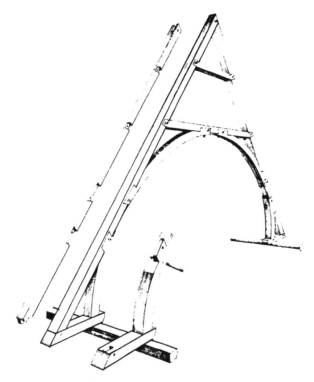

Fig. 96. Greyfriars' church at Lincoln, roof construction.

as a result ordered the sheriff of Essex to pay for repairs and repainting in the chapel of St. John in the churchyard there . . . In 1258 the Oxford Parliament forced William of Valence to leave England, and the manor of Clavering was apparently in the king's hands until Robert came of age in 1265. From 1297 he was Captain of the Marches in Northumberland, but surprisingly he took his surname from his small Essex property. In 1301 he signed the Remonstrance from the barons to Pope Boniface VIII as "Robert fitzRobert, lord of Clavering", while from 1297 onwards his eldest son John was known as "John de Clavering". Robert's six younger sons also used the same surname. According to the cartulary of Sibton Abbey, the surname was adopted at the instance of Edward I, because it was Robert's principal manor. At Robert's death in 1309 the Inquisition found "a certain messuage with easements of buildings and fruit of the garden worth 6s (8d?)": the earliest documentary reference to the Bury, although there was presumably a manor house at Clavering when Henry III stayed there in 1251.'

This building still contains a long open and aisled hall with a spere-frame and scroll-moulded capitals, a storeyed service end, and strong evidence for a former solar end that has been demolished. The framing is illustrated in Fig. 95. Long scissored passing braces were used in each cross wall together with notched-lap-joints with refined entry angles, and the service-bay first floor is lodged. At the former solar end there is a top-plate scarf of high quality, stop-splayed and tabled with under-squinted and sallied butts. The structural details suggest a date during the early 13th century—which, if proven, would extend the generally accepted period for the use of scroll mouldings, which is currently thought to begin later.

The Greyfriars' Church, Lincoln

The Grey Friars began building their chapel in Lincoln during the first half of the 13th century, and possibly before 1237 (*Archaeological Journal*, 1935, 42–63). This build has a 21½ft. span, and a length of 72ft; the total length today is 101ft., owing to two bays of nave which were added to the west end, probably at a date close to *c.* 1260. The timber roofs built over these two parts are dissimilar, and the earliest (at the eastern end) provides an example of the use of 'compass' timber, producing a semicircular archivolt. A part of this design is illustrated in Fig. 96. The ends of the collars are notch-lapped, but the compassed soulaces and ashlar pieces are not; they have subtly modified lap dovetails similar to those noted at Tewkesbury Abbey.

The Old Deanery, Salisbury, Wiltshire

The top plates of this building were jointed by means of what may be the first known example of stop-splayed scarfs having under-squinted abutments, together with tonguing and grooving, which in these examples is a single tongue running through both of the tablings. This joint is illustrated in Fig. 250, and I am indebted to the Royal Commission for information concerning its interior details. The building, and therefore the scarf, are datable by reference to a published deed of gift issued by Richard de Wykehampton, who became Bishop of Sarum in 1258 (C. R. Everett, 1944, 425). This deed states: 'we assign, grant and give forever those our houses in the Close of Sarum, which we were accustomed to inhabit when Dean'. This has been taken to indicate a date of construction between 1258 and 1274 (N. Drinkwater, 1964, 44).

Wynter's Armourie, Magdalen Laver, Essex

Today this is an H-planned house, but this may not have been its original form. It occupies a moated site in a parish without a nucleated village which has been almost completely cleared from what was once an extension of Epping Forest, by means of linking up assarts based on a number of similar tenements, often moated. This is the completed 'Wealden' situation, and much of the work was done in the later Middle Ages, from the 13th century onwards. An Alan Wynter was in occupation by 1248, and his successors continued to prosper, unhampered by much in the way of manorial incidents (*V.C.H.*, 1956, 193). The framing of the hall of this house is shown in Fig. 97, illustrating the main curiosity of the building—the low tie beam which crosses the hall at little above head level. This indicates that the floor has been raised and is strictly comparable to Gatehouse Farmhouse (Fig. 116). There is evidence for the former existence of a crown post on the central collar. Such mouldings as exist are inadequate for dating purposes, but the scarfing of the 'purlins' is more useful in this respect. These joints are stop-splayed with tables and under-squinted abutments, feather-wedged, and with no less than ten face pegs. A date of *c.* 1290 is suggested, and slightly earlier dates are not precluded. The solar wing is of archaic framing with widely-spaced studs, sole bracing, and brace-supported jettying (Fig. 98). This may be contemporary, and sustains comparisons with the jetties of Priory Place.

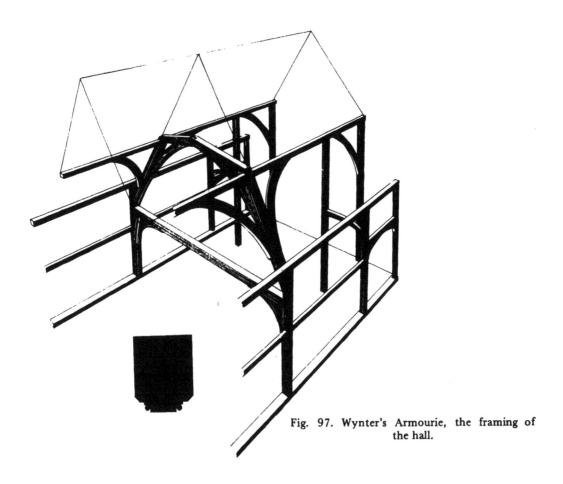

Fig. 97. Wynter's Armourie, the framing of
the hall.

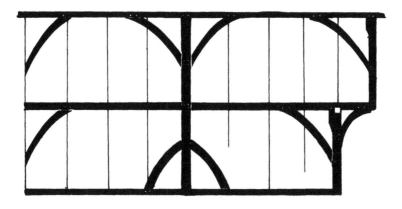

Fig. 98. Wynter's Armourie, wall frame of solar wing.

The High-Roof above the North Transept, Westminster Abbey, London

Documentary evidence relating to the completion date of the three eastern arms of the abbey has been published (H. M. Colvin, 1963, 130), and 1269 is most probable; the master carpenter to the king, Master Alexander, directed the work. The roof comprised 36 frames, each with two collars and scissor braces, the feet of the latter being fitted into secret notched-lap-joints. The tie beams were placed at intervals of eight couples inclusively, and were of the distinctive type with branching ends. The ashlar pieces were in-canted. Perhaps the most interesting feature of the roof was the fitting of diagonal braces into trenches on the outer faces of the rafters, at the ends of its range; these were to inhibit the racking of the couples. A short section of this is illustrated in Fig. 99. Nothing remains today, because all but the eastern apse was replaced shortly after 1964.

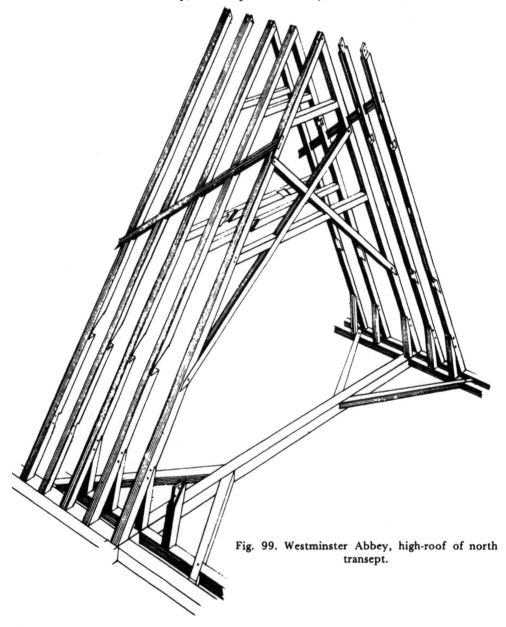

Fig. 99. Westminster Abbey, high-roof of north transept.

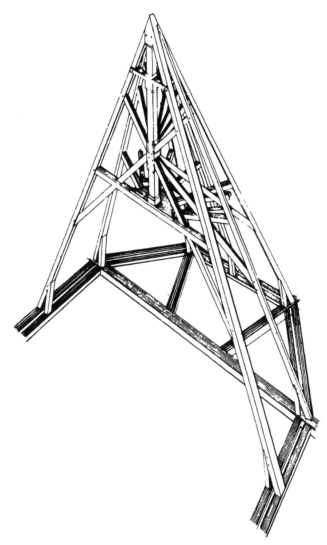

Fig. 100. Westminster Abbey, high-roof of eastern apse.

The High-Roof above the Eastern Apse, Westminster Abbey, London

The Westminster apse roof was entirely dismantled and then reframed, any defective timbers being treated or restored as deemed necessary, soon after the transepts had been re-roofed. The eaves plan over which it was built was basically a five-sided one, and the span was reduced between two tie beams, with the high king post mounted halfway between the two. Against the king post rest five virtually lean-to roof frames with half collars and vertically halved scissor braces, all tenoning into the thicker parts of the post. The rafters in the facets of the apse rise in-plumb and parallel, with tapered tops spiked to their converging neighbours, as in the case on the facets of the chapter house spire roof at York. A perspective of this roof is shown in Fig. 100. This arm would have been roofed earlier than the transepts mentioned above.

113

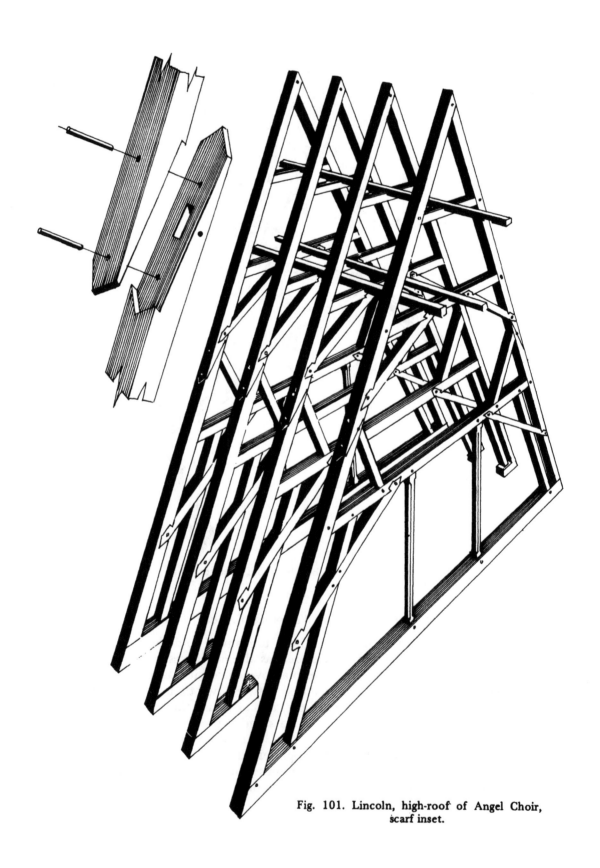

Fig. 101. Lincoln, high-roof of Angel Choir, scarf inset.

The High-Roof of the Angel Choir, Lincoln Cathedral

The Angel Choir was built between 1256 and 1280 by the architect Simon of Thirsk (J. H. Harvey, 1961, 141). It retains its timber roof, which must have been complete long before the final date in order that the vaults could be built indoors, as was both customary and essential. A bay of the carpentry is illustrated in Fig. 101, which shows the relationship with the roof of the nave east of the crossing tower; the slender and octagonally sectioned queen posts were fitted as collar supports, and open notched-lap-joints were still used for the soulaces. Oaks long enough to form the rafters were not available at this date, as is evident from the fact that most were scarfed, as illustrated, with splayed joints having taced and under-squinted abutments. The tie beams were still placed at intervals of every fourth couple, and the most significant advance was made in providing anti-racking members which took the form of *three* collar purlins, each with its soffit trenched to house the faces of the collars.

The North Door Leaf, Church of St. Mary, Stifford, Essex

There is no reason to doubt that this door leaf is not original for its position, and that it compares favourably with others of the late 13th century; the ironwork is applied to planks of oak which are both humped and v-edged. The rear frame is simple, comprising eight cross ledges with a harr and a head, the spikes being clenched—that is, having their points re-turned and driven into the ledges. This and other similar doors, such as the great north door of Durham Cathedral and the south door of Heybridge Church in Essex, indicate that from the earliest times those craftsmen who used only timber were a minority of 'pure' carpenters, whilst others were content to use iron and timber in equal measure (Fig. 102).

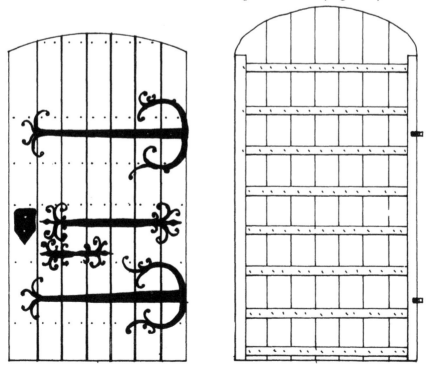

Fig. 102. Door leaf from Stifford church.

Framed Vaults of the Presbytery, St. Albans Cathedral, Hertfordshire

The Presbytery was built between 1235 and 1290 (J. H. Harvey, 1961, 155). It seems that the original intention to vault in stone, for which evidence may be seen (H.M.S.O., 1967, 14), was never implemented, and that the existing timber-framed vaults are the originals. These are probably the earliest framed vaults in existence. A section is illustrated as Fig. 103, and a cross section of a rib as Fig. 356. The reason for the choice of timber seems to have been economy. The pattern of the ribbing was altered from a lierne pattern by Wheathampstead to the existing one of surface ribs from the springers to the tranverse ribs. In his time the bosses were moved to their present positions. In 1680-1, according to the latter source, 'new backbones were given to the lower ends of some of the ribs. The present shields which mask the junction of the stone springers and the wooden ribs date from this repair . . .'. In 1930 it was found to be endangered by worm, and was reinforced with steel. That this framed vault was formerly integral to the high-roof is clear, but no details of either the high-roof or its integration with the vaulting have survived.

Fig. 103. Framed vault of St. Alban's Presbytery.

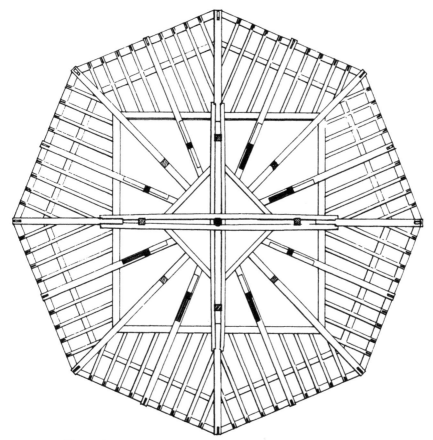

Fig. 104. York Minster, plan of chapter house spire at base.

The Chapter House Spire, York Minster

This is among the finest surviving works of the Gothic carpenters of the great age. The 'house' itself is dated 1285–96 by Dr. John Harvey (J. H. Harvey, 1961, 169), but its carcase of masonry has been altered and its height increased at a time when a trabeation of stone was built between the upper parts of its buttresses and the returns of its plan, which is octagonal. The timber roof, in the form of an obtuse spire, is of uncertain date since it shows a doubling of framing at eaves level. Whether it dates from before or after the raising, or partially from each phase, is open to question. The span inside is 58ft. clear and this was bridged by two 'built' transom beams, each formed of three timbers. A plan of the work at the eaves level constitutes Fig. 104. The longest timbers were forcibly 'pinched' round the central mast, trapping extending end timbers, as illustrated; their tendency to spring straight again was prevented by trapping in housings at the edge of the square void. In the plan (Fig. 104), the ring of alternately blacked or hatched squares on the radially disposed 'floor' timbers represents the cross sections of the vertical timbers, these either suspend the timber-framed vaults, or rise between the spire's second set of pinching collars. The complexities of the upper framing are shown in Fig. 105.

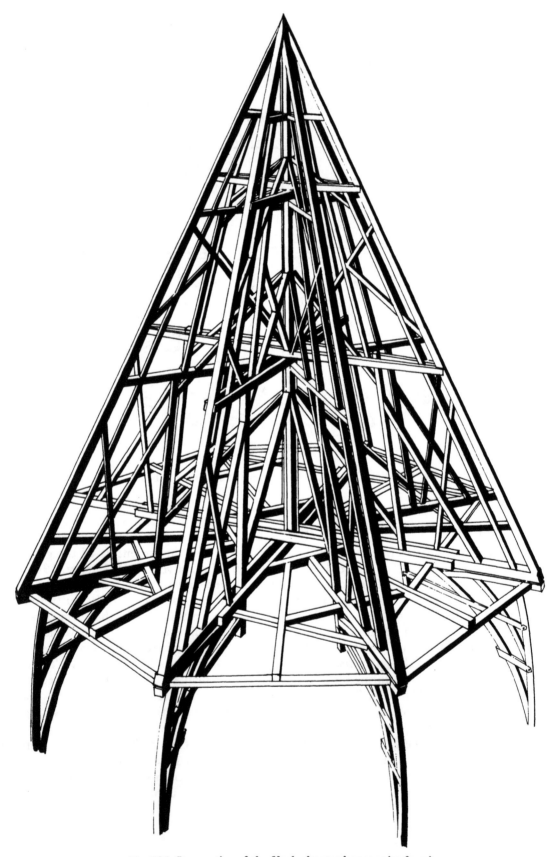

Fig. 105. Perspective of the York chapter house spire framing.

Fig. 106 shows the pattern of the framed vault; it is of the type without a central pier, as at Wells, Southwell and Old St. Paul's (J. H. Harvey, 1961, 186). The octagonal timber-framed cone is drawn with enough timbers omitted to clarify its construction. Its height is about sixty-four feet and the spire mast is scarfed where necessary, the joint used being stop-splayed with tongue and groove, edge-pegged, and worked upon an octagonal section. This joint relates to others of the reign of Edward I. The problems of assembling and rearing this great frame are rather similar to those solved at Ely later, when the lantern was raised above the octagon. It is highly probable that the masons' scaffolding, which must have been both external and internal, and of full height, was taken over by the carpenters; after which the same problem of simultaneously fixing the eight 'wall brackets', which are all framed with open notched-lap-joints, occurred. Theoretically, these were locked immovably after the timbers of the internal square void and the external octagonal wall plates were positioned: it is suggested that eight carpenters may have been used concurrently during this difficult operation, as was certainly recorded at Ely.

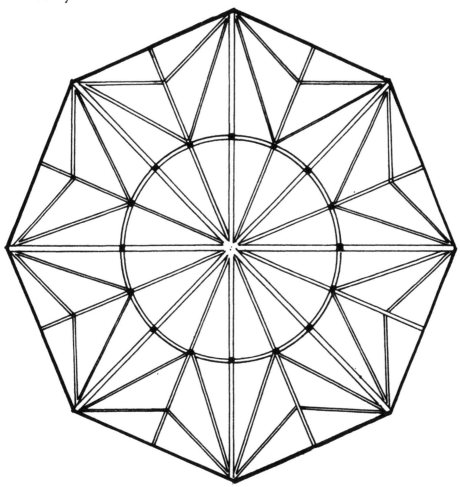

Fig. 106. York chapter house spire, pattern of the vault ribs.

Such great complexes of framing as these become extremely stable once assembled, but are not so until complete, and their synchronous assembly is difficult to envisage although it was quite obviously achieved. The wall plates have pairs of single tenons at their joints with the radial hammer-type beams, and the notches cut into the lap joints are of the 'archaic' profile; a selection of structural details is illustrated in Fig. 107. The disposition of the timbers within the cone distributes the weight equally between the eight bracket frames set in the angles, then through them against the external buttresses.

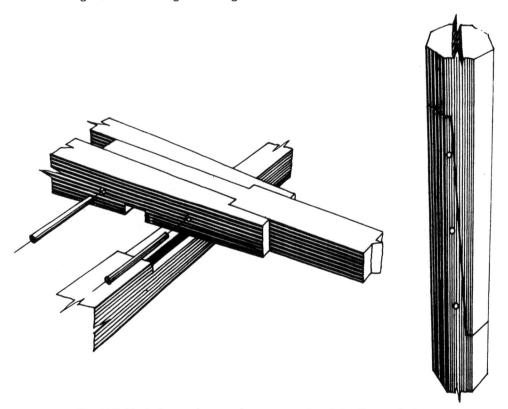

Fig. 107. York chapter house spire, constructional details, scarf of mast.

The Apse Roof, Church of St. Nicholas, Little Braxted, Essex

A small Norman church, with nave and eastern apsidal chancel, of uncertain date, as is the apse roof. This is illustrated in Fig. 108, with a view of its remarkable compassed collar beam isolated to the left of the drawing. The rafters are scissor-braced without collars, except for the couple placed on the tie beam which is diametrical to the semicircle concerned; this couple also has a high collar, on which the radial rafters could be lodged to make the semi-conoid form. The method used to effect their convergence is shown at lower right of the illustration. The scissors' feet are chase-tenoned, which suggests a relatively late date during the 13th century; the through-splayed scarf joint used to form the curved collar precludes a later date, however.

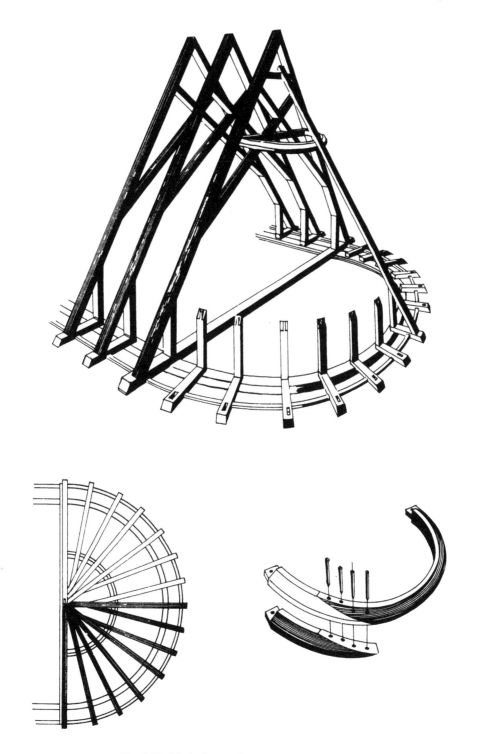

Fig. 108. Little Braxted, construction of apse roof.

Place House, Ware, Hertfordshire

This aisled timber-framed hall is now encased in 16th- and 17th-century work, and is sited on the north-east fringe of the old town. From the 1660s to the 1760s it was used by Christ's Hospital as an overflow to their buildings at Hertford. It was not among the original endowments of the Hospital (J. How MS., 1582), but was acquired in 1685, after some years of use, from one William Collett (E. M. Hunt, 1949, 10). It had apparently been sold in 1575 with what remained of the original manor of Ware to the Fanshawes (*V.C.H.*, 1912, 388). In about 1667 the lands were broken up, and this was presumably when Collett got his parcel. Much of the original manor had been granted before 1080 in frankalmoign to the alien Priory of Ware (*V.C.H.*, 1914, 455), a dependency of St. Évroul and not to be confused with the local Franciscan friary, since the Greyfriars of London (on whose site

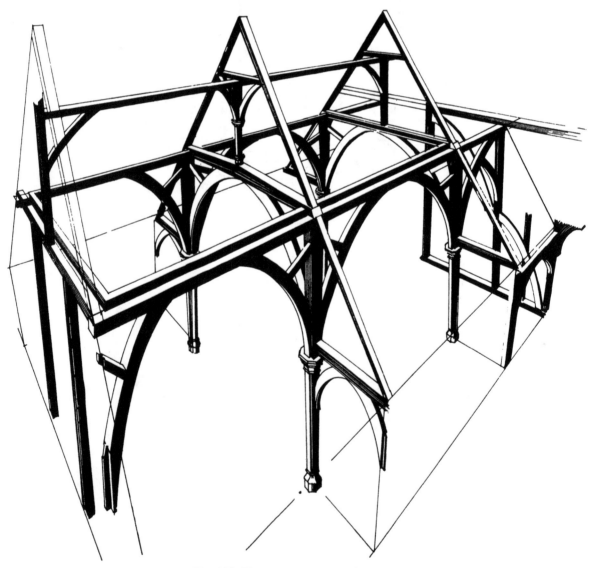

Fig. 109. Place House at Ware, framing.

122

Christ's Hospital was established) had no power over their brethren in Ware, and the lands did not pass together. The endowments of the alien priory were granted to the Carthusians of Sheen and afterwards to Trinity College, Cambridge. Until the former dissolution (1414) the patronage had remained with the lords of the lay part of the manor, who in the 13th century had built a lay residence for their own convenience on priory land and near the conventual buildings. The location of the priory seems to have been established: the site, just north of the church which has been suggested (*Trans. E. Herts. Soc.*, 1906, 119-32), is more probably that of the rectory which they held.

'Place House' is an appropriate name for a spacious residence on ecclesiastical property and it seems probable that it represents, in some fashion, the house built on the priory precinct. It seems too late for the original structure, the complaints about which name the Countess of Winchester, before 1235, and Joan de Bohun. After the latter's death in 1283 steps were taken to demolish it, but in 1295, soon after her nephew and heir had taken livery in 1290, the priory was in royal hands. It was perhaps at this time that a compromise was made which allowed the erection of a new house on or near the same site and henceforth attached to the lay manor— on which, in 1515, two mansions were recorded, the Bury and Kiddswell (*V.C.H.*, 1912 387).

The frame of the house (Fig. 109) is among the most sophisticated and finely wrought known, and comprises a two-bay open hall with aisles, a cross entry, and a jettied service wing: it may well have suffered the loss of a solar wing, possibly demolished to allow for the creation of the existing thoroughfare. The timbers are richly moulded, as shown in Fig. 357, which indicates the sections and profiles of the principal parts. These profiles suggest a date of *c.* 1295 (H. Forrester, letter to author), which is consistent with the hypothesis previously given, and suggests that a new house was erected some twelve years after the demolition of the original intruded house.

St. Etheldreda's, Ely Place, London

John de Kirkby became Treasurer of England in 1284 and Bishop of Ely in 1286, he died in 1290, without having built the oratory for which his predecessor had obtained a licence from the Dean and Chapter of St. Paul's in 1251. It seems that his successor, William de Luda, may have been responsible for the existence of the church on this site today, which was the official London residence of the Bishops of Ely.

As in several other cases, documentation exists for this building, but it is of no help in determining its date. According to the guide printed for the church, a document indicates that it was finished by 1299. However, there is no certainty concerning the age of the most interesting feature, the timber floor of the nave, which is also the ceiling of the undercroft. This is the largest area of timber flooring at present known that is mounted upon samson posts (the floors in the White Tower are here discounted, being altered beyond recognition). There are six samson posts and a timber respond against the west wall, which now stand upon stone columns; but examination proves them to have been in their original state prior to a restoration in the 19th century. Each post supports a bolster, aligned along the crypt, with its ends cut to double-hollow profiles; along these the bridging joist is laid. There are 50 common joists (each 10ins. square and laid 5ins. apart) lodged across this, some being half the necessary length, and butted on top of the bridging joist. The posts are braced in four directions, as shown in Fig. 110, and above the third post from the west there is the scarf shown in Fig. 243, which is stop-splayed with countered tongues and grooves.

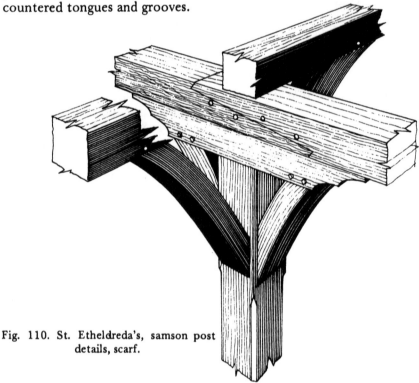

Fig. 110. St. Etheldreda's, samson post
details, scarf.

The Little Hall, Front Quad, Merton College, Oxford

The College records concerning this building indicate that it stands on the site of the first three houses bought by Walter de Merton for his new foundation between February 1267 and the end of 1268. At the date of acquisition the building was known as Flixthorpe's House, being owned by one Robert Flixthorpe, before which its existence in some form can be traced back to 1264. It seems that little remains from the earliest dates other than masonry in the carcase walls, but the Rolls of Merton record much building activity between 1299 and January 1300. At this time it became known as the Warden's House, or *aula custodis*. It is to this period of rebuilding that the timber roof can be attributed (Dr. J. R. L. Highfield, 1970, 14–22). The hall measured a little over 28ft. and spanned 21ft., its stone walls having restored traceried windows on the south which also date to the same rebuilding operation. The roof had one central and two terminal tie beams, slightly cambered, mounting queen posts with capitals and bases, all of which were set upon wall pieces with arched braces. The side purlins were set 'flat' and tied by higher tie beams, upon which a crown-post and collar-purlin assembly was mounted. The roof, as restored in 1969, is illustrated in Plate III, and one queen post in Plate IV.

124

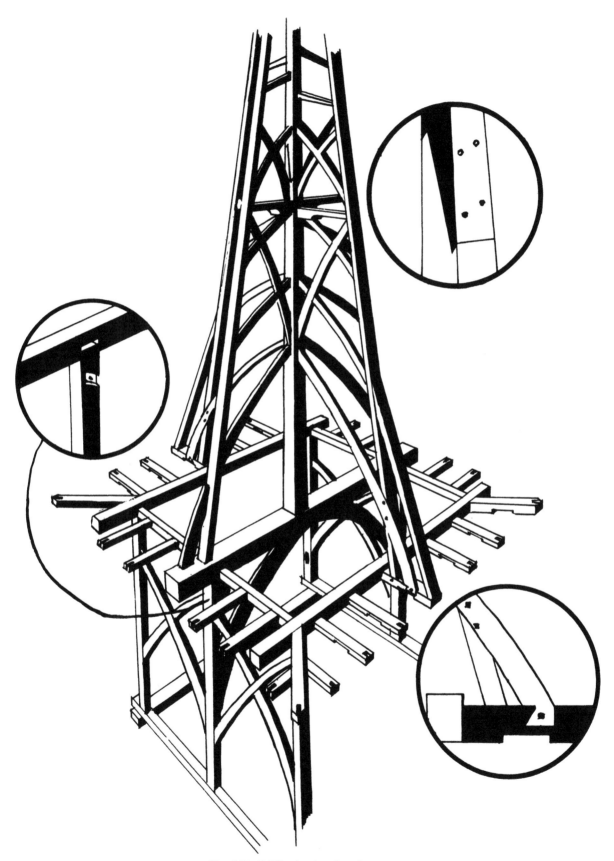

Fig. 111. Stifford spire, framing.

The Church of St. Mary, Stifford, Essex

The west tower of this church is generally ascribed to the 13th century, and the surmounting brooch spire is contemporary with it. The top stage of the tower is fitted with a saltire-braced timber wall frame on the north and south sides, and the central posts of each wall have jowls and arch braces to the heavy transom beam on which the spire mast is stepped; in this respect, and as viewed from east or west, this stage resembles normal domestic timber framing, like that of Priory Place, and is contemporary with it. The use of the jowl also endorses a date ascription to the end of the 13th century. The saltire braces are lap-jointed without any notching and indicate the order of assembly for the wall frames, being chase-tenoned into the centre posts. The wall plates of the spire are mounted on the wall posts, their soffits level with the upper arrises of the masonry; the sole pieces tenon into them, and diagonal sole pieces with jack sole pieces radiating from them provide for the returns under the brooches. All the sole pieces are trenched on their soffits to fit over a partly embedded mid-wall plate (see Fig. 111).

The spire mast is scarfed, as shown in the inset drawing; the lap joints of the wall frames and the trenching on the soles' soffits are also inset. The spire was framed with delicately curved braces, the first four of which pass through squint trenches on the cardinal rafters, above which an elegantly interlaced sequence rises just above the collars (which were also jowled). Above the collars are four pairs of saltires, and from there to the apex are down-canted struts (as at Canterbury Cathedral spirelet); a date between *c*. 1270 and *c*. 1300 is proposed for this work. The simple and compressive splayed scarf used for the spire mast (inset) closely resembles that in the spire mast of the Rhenish helm at Sompting.

The Service Wing, Tiptoft's Hall, Wimbish, Essex

Today this is a large H-plan house on a moated site situated in an area of scattered farmsteads, with few integrated villages, and it is close to both the parish church and the capital manor-house. The holding was reckoned to be a second manor from the 14th century, and presumably takes its name from the occupancy of Sir John Tiptoft between 1348 and 1367; but it was also occasionally known as Wantone's (P. Morant, 1768, 558). Since neither the period of Tiptoft's occupation nor the occupation by Wantone before him relate to the main periods of expensive building activity, however, previous wealthy owners obviously existed. One or other of these knights may have been associated with part of the rebuilding; however, Wantone died in 1347 possessed only of this house. The most ancient part of the existing house is the service wing, illustrated in Fig. 112, wherein only the frame is shown: it embodies several features that place it among the few demonstrably experimental examples of jettied timber buildings.

The house is now partly encased in red brick, giving little external indication of its long and complicated history, which postulates either a T-plan or an H-plan house on the site before the existing hall and solar wing were built. The service wing has, in two of its storey posts, the empty mortises for the top plates of a former and contemporary hall. The first floor has been repaired in places but seems to have been half lodged, and then half framed into the lodged half; while the

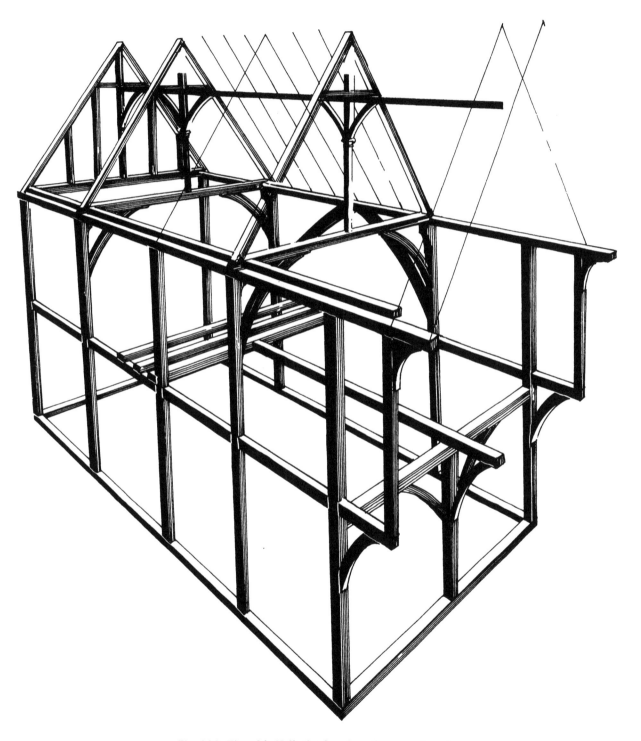

Fig. 112. Tiptoft's Hall, the framing of the service wing.

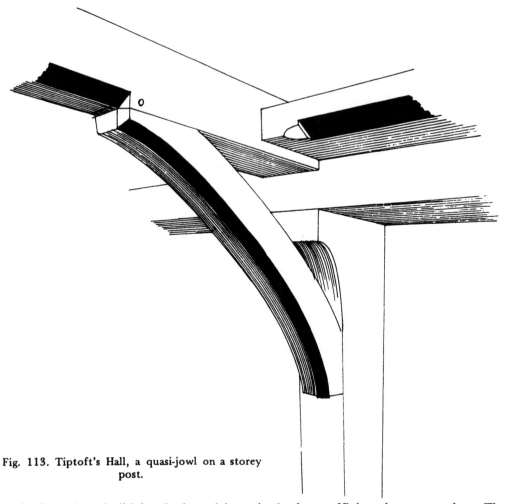

Fig. 113. Tiptoft's Hall, a quasi-jowl on a storey
post.

method used to build its single end jetty is the least efficient known to date. The
storey posts have the archaic features of thickened tops which are not jowls, but
which visually resemble jowls (Fig. 113). This type of thickened post head also
exists in the belfry at Bradwell-juxta-Coggeshall (Fig. 47), and in the belfry at
Aythorpe Roding (Fig. 86), and it is probable that some examples pre-date the use
of the upstand for the triple joint eventually adopted at these points.

The roof has what appear to be king posts, but despite the fact that they have
butted and tenoned collar purlins between them they do not reach the rafters'
apexes, except at the end gables. The capitals of these have accurately cut and
very rare mouldings, namely pointed rolls, which were repeated upon the capitals
of the four carved and partly octagonally sectioned jamb posts of the three service
doors; these mouldings are discussed under that heading and shown in Fig. 329.
The entire wall framing is archaic, being the most widely spaced known to the
writer; and, instead of studs, the ground-storey wall beneath the jetty has a samson
post placed almost centrally in its span, with complete evidence for the three
original braces.

A large hall-house which is possibly among the first in England to presage the H-plan. It was built roughly parallel to, and about thirty yards west of, the site of the western claustral range of the once large and wealthy Augustinian priory. This priory was founded by Geoffrey Baynard in 1106, and judging by the surviving fragment (now the parish church), it was sufficiently prosperous to undertake expensive building operations early in the 13th century (*V.C.H.*, 1907, 150–4). The house may have served as a guest hall for important patrons (*cf.* Bricett and, perhaps, Kersey—C. A. Hewett, 1976, 47 and 48), but was strictly extra-claustral.

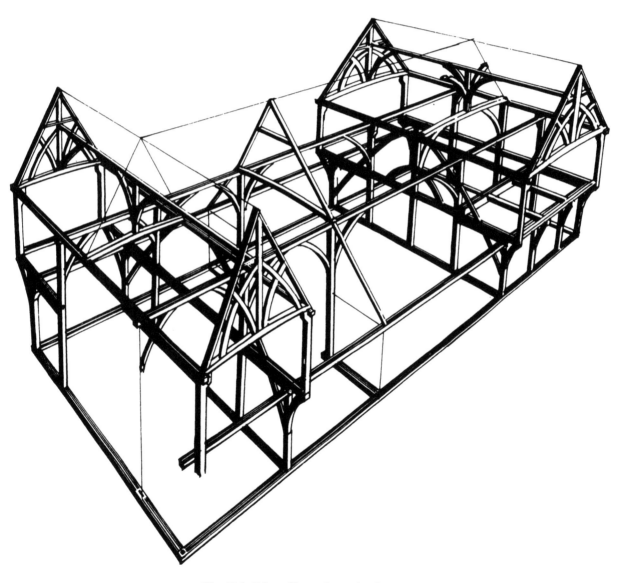

Fig. 114. Priory Place, the entire frame.

The whole structure was renovated some years ago, when it could be examined in complete detail; it was indubitably of one build and originally comprised an open hall with jettied wings of different spans at either end. The whole frame is shown in Fig. 114. The ground plan was a plain rectangle and the first-floor plan H-shaped due to the jetties, of which there were four. The central third of the first-floor wings was lodged, and the other two-thirds were framed. The most obvious archaism of the jetties is that none of the over-sailing joists rests on the ground-storey wall (Fig. 115). The majority were instead supported on braces from the studs. The side girts were doubled to make the jetties, an upper one lying on each and projecting a short distance beyond it; a large quantity of timber was therefore used to achieve a minimal projection. The jointing of all tie beams was by lap dovetails with entrant shoulders, the scarfing by means of stopped splays, and the roofs were crown-posted octagonally and plainly. Saltire braces were fitted into each gable and the walls were entirely free of wind bracing, with their studs widely spaced. Only two moulded timbers survived in the house, the more significant of which was a scroll-moulded length affixed as the high-end dado.

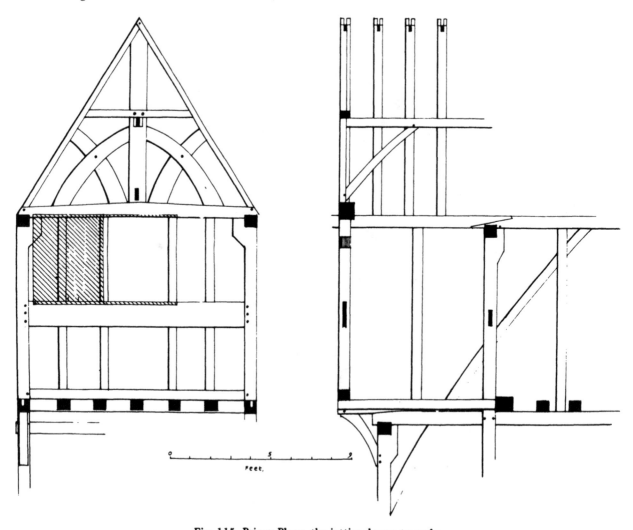

Fig. 115. Priory Place, the jetties drawn to scale.

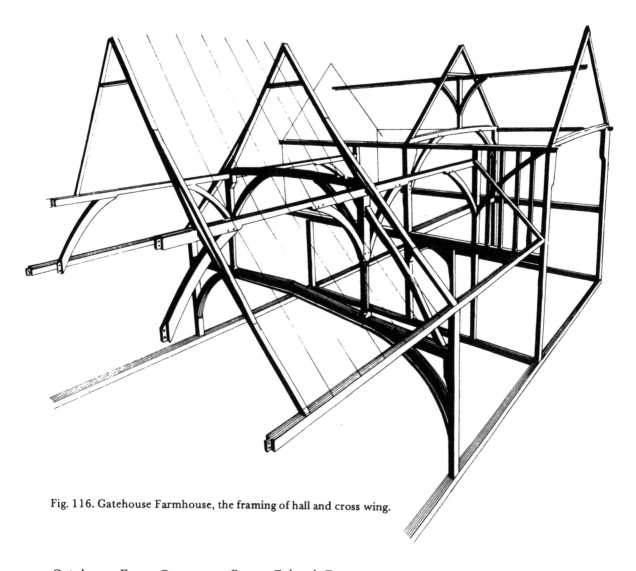

Fig. 116. Gatehouse Farmhouse, the framing of hall and cross wing.

Gatehouse Farm, Gransmore Green, Felsted, Essex

When examined by the Royal Commission early in the present century, this was an H-plan house with a central hall open to its ridge; one cross wing has since been demolished, and the tie beam crossing the hall cut away to produce a very unstable and crude visual semblance of two hammerbeams. The service wing retains one four-centred timber door head, and its floor joists are fitted by barefaced soffit tenons, facts which together indicate an early 14th-century date; but the elaborate mouldings of the two queen posts and their water-holding bases support an earlier dating. The entire frame is illustrated in Fig. 116, in which the tie beam has been hypothetically drawn in its original entirety; and a queen post's mouldings with a cross section through the tie beam are drawn as Fig. 326. As to the complexity of dating, it is considered that well cut water-holding bases are not known after *c.* 1300 in any context (H. Forrester, 1972, 28–29); and the dual incidence of the roll with frontal fillet between 1220 and 1300 and also between 1400 and 1500 admits of only the earliest interpretation in this case. The service wing is therefore considered to be a 14th-century addition to an originally rectangularly planned great hall, which in view of its superlative quality, was probably of monastic origin.

131

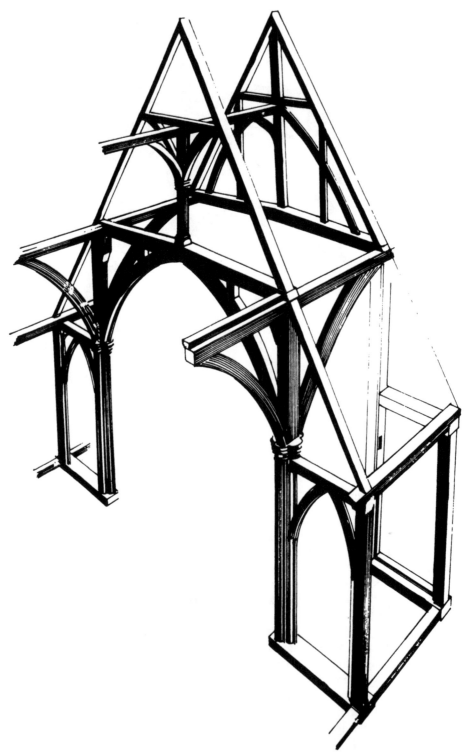

Fig. 117a. Little Chesterford Manor House, framing.

132

Little Chesterford Manor House, Essex

A capital manor-house immediately adjoining the parish church in an area of early and intensive settlement with easy communications. During the 13th and 14th centuries it was sub-infeudated to successive generations of the le Breton family as a single knght's-fee (P. Morant, 1768, 556–7). The house has developed into an H-plan shape, the earliest part being the service wing, which was originally a stone-built upper hall with ground-storey services; to this was added the aisled timber hall, and soon afterwards the timber solar wing. It has been thought that the whole building was completed by *c.* 1330 at the latest. This is unlikely because the scarf used for its top plates is identical to that used for Cressing wheat barn in *c.* 1250, and an earlier date is now regarded as more probable. The two frames from the lower end of the hall, together with the entire timber frame, are shown as Fig. 117, and a section through an arcade post and its capital as Fig. 330.

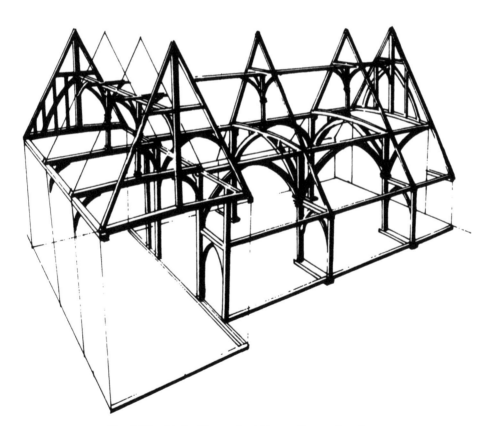

Fig. 117b. Little Chesterford Manor House, framing.

Southchurch Hall, Southchurch, Essex

This is a large manor-house on a moated site some distance from the church (both are now engulfed in suburbia). It was held of Christ Church, Canterbury, by an eponymous family the last of whom, Peter de Southchurch, died in 1309, when his heiress conveyed it to William de Flete (P. Morant, 1768, 298). The manor was a large one with 640 acres of arable land, mills, money-rents, and so on, and however much Christ Church drew from it, its occupier could afford to build this internally showy house at any date close to that cited. The first of the surviving structures was an open hall, of considerable span and without aisles, the framing of which is illustrated in Fig. 118, from which a later and jettied cross wing is omitted. The gable frame at the service end incorporates crooked timbers that may be 'eaves blades' or full-height cruck blades, but this question cannot be resolved, due to the thorough cladding inside and out. If they are the former then many parallels exist in Kent and Surrey: Littleton, Artington, Elstead, and Etchinghill, Benover and Stockbury. All the scarfing is through-splayed and tabled without wedges, since no abutments exist, except at the centres of length. One possibly unique feature is the soffit cusping along the archivolt of the tie beam, which is partly carved and partly applied; on this beam is mounted an elaborate crown post (Fig. 333).

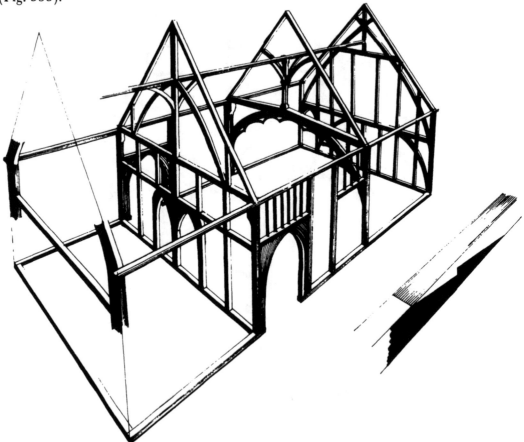

Fig. 118. Southchurch Hall, the framing, with enlarged scarf.

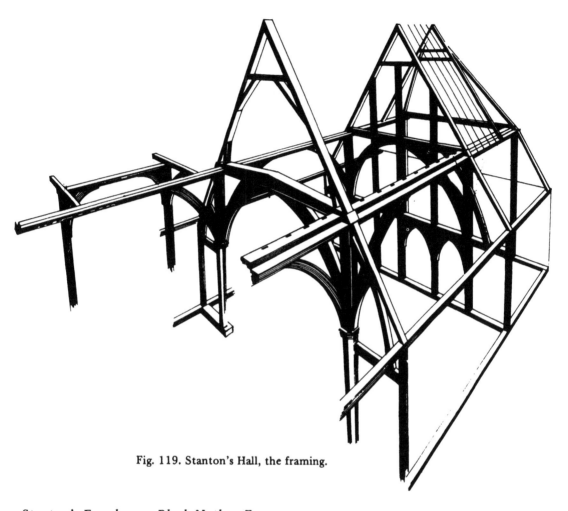

Fig. 119. Stanton's Hall, the framing.

Stanton's Farmhouse, Black Notley, Essex.

The surviving documentation concerning this remarkably fine building indicates a very early date: '34 Edward I, 1305-6. 869 Mich. Thomas de Staunton and Mabel his wife, pl. Giles de Lenham and Alice his wife, def. I messuage, I carucate of land, 6 acres of wood, and 4 acres of meadow in Black Nottle and White Nottle. Plea of covenant. Thomas acknowledged the premises to be the right of Alice, and for this def. granted the same pl. and the heirs of the body of Thomas, with the remainder to Reginald, son of Roger de Bockyngg and Thomas, brother of Reginald, and the heirs of Reginald'. (*Essex Archaeol. Soc.*, 1913, 112.) The mouldings carved on the timbers within the formerly open and aisled hall suggest a date no later than *c.* 1300. The ground plan was rectangular, providing two central bays of hall, and service—and solar bays at either end. A part of the service bay which survives has three finely-cut door heads with blind tracery, the central one formerly leading via a passage (for which structural evidence exists) out of the house, doubtless to a detached kitchen building. A major fragment of the front cross-entry door head also survives, with blind tracery of the highest quality. The general framing is illustrated as Fig. 119, and the mouldings and decorative details as Fig. 332.

135

The floor joists at first-floor level in the service bay are of interest in that they are fitted with unrefined tenons that were cut a little nearer to the soffit than the upper face; this presages the introduction of true soffit tenons a few years later. (A re-examination during the preparation of this text has revealed that unfortunately restoration has either destroyed this evidence, or concealed it; but the fact is here recorded, as are the main-post base mouldings that were reburied when the floor was remade.) The arched braces spanning the great hall are worked with precision into a roll with three fillets, divided by two almost circular hollows with a fillet each side; this profile is worked round the very long curves of the arches involved. It is not known how this was done, and no hand-tool that could produce this moulding is known to survive. According to Forrester (1972, 31), this profile dates between c. 1270 and 1330.

Wells Cathedral, Somerset

The chapter house at Wells was completed by 1306 (L. S. Colchester, paper for Newman Soc., Bath, 1971), and the event was recorded in Charter 165: *Propter magnos sumptus et expensas quos insi aposuerunt atque fecerunt circa construc-tionem sui capituli,* work at plinth level having been resumed in 1286. The roof must therefore date from the final year of building, since it stands on the vaults, and could not have been positioned until the rest was complete. The building is of octagonal plan with a central pier. The timber roof continues this central pier as a post; has an intermediate octagonal sill (also of scarfed and angled timbers) which stands upon the crowns of the vaults; and eight wall posts, each on a timber sill placed in the internal angle of the plan. By this means eight radiating principal-rafters are supported, together with seven sets of tenoned, concentric, octagonally disposed purlins, upon which the common rafters were laid parallel and at right angles to the octagonal sides of the carcase. This roof is illustrated in Fig. 120, a general perspective from which many timbers are omitted for the sake of clarity. The scarfing of the timbers forming the octagonal sill is of interest, being through-splayed with counter tongues and grooves, and edge-pegged. These joints are placed on the straight lengths in order that the angles may be solid, strong timber, and are shown in Fig. 257, which shows how the angles were so clipped as to be parallel to the sides of the posts' mortises. The jointing of the octagonal purlin lying immediately above the intermediate posts is thorough, and highly complex; the posts are mitred beneath the purlins and fitted into them by pairs of tenons. It is not possible to see the form of the posts' tenon, which must impale this assembly.

The Choir Stalls, Winchester Cathedral, Hampshire

In 1308 the Bishop of Winchester requested from the Bishop of Norwich the protracted services of William Lyngwode, carpenter (a request that was repeated in 1309), in order that he might finish the construction of the stalls (S. Jervis, 1976, 19). The higher, rear floor of these is fitted by what are at present the earliest known examples of barefaced soffit tenons (C. G. Wilson, 1976, 16-18). (Fig. 292.)

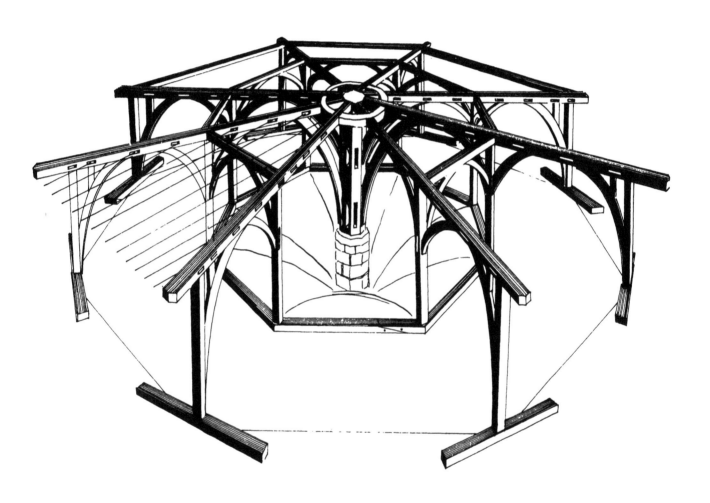

Fig. 120. Wells Cathedral, roof of the chapter house.

Fig. 121. Rochester Cathedral, roof of the south choir.

The Roof of the South Choir Aisle, Rochester Cathedral, Kent

This roof was dated by St. John Hope (1900, 82) to shortly before 1322. It is a lean-to roof of very shallow inclination (Fig. 121). It was designed to stand on its own wall posts, which were mounted on carved timber corbels, and support an interesting and truncated tunnel-type framed vault, which is today plastered between the moulded and painted rib timbers. Traces of original colour were recorded when one rib (a section of which is shown in Fig. 357) was last examined; red for the rounds, green for the hollows, and white for the fillets.

The Prior's Chamber, Prittlewell Priory, Essex

Prittlewell Priory 'was founded from Lewes, the chief Benedictine house in England, in 1121' (Sir N. Pevsner, 1956, 322), and this building may well have been what its present name implies, containing as it did a heated open hall upon its first floor. It was a long-wall jettied house with a crown-post roof, which survives, as does clear structural evidence for the mounting of its 'fumer' above the apparent hearth position. The framing above first-floor level is illustrated as Fig. 122. Several structural features are of interest, such as the fitting (by means of a rebate on the top plates) of an internal ashlar plate, which dispensed with the usual sole pieces for roofs of seven cants; and strutting, to stiffen the arch braces, placed within their open spandrels. The wind bracing of the walls is effected entirely from their base lines, in most cases the timber faces of the jetties. The cross-entry doorways are actually the ends of a stone tunnel, the flat top of which provided a hearth for the fire.

Fig. 122. Prittlewell Priory, framing of the Prior's Chamber.

Baythorne Hall, Birdbrook, Essex

This house, and not the one named Baythorne Park to the east of it, appears to represent the original capital messuage of the Baythorne estate. It lies close to the River Stour and is exposed to flooding in wet winters, which suggests a change in the flood areas of that river since medieval times, since sites prone to occasional inundation were then, as now, avoided. The house is isolated, but has a secondary settlement between itself and the other house. During the 14th century it was held by the Wantones, as was Tiptofts at Wimbish, and it remained with this family until 1391 when the estate was divided betwixt coheiresses. It is possible that this hall was built by a Wantone at the time that the Wimbish manor had been released to Tiptoft (P. Morant, 1768, 344).

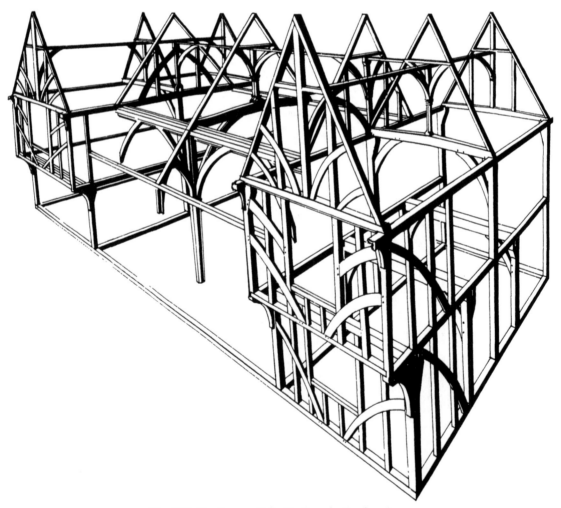

Fig. 123. Baythorne Hall, Birdbrook, the framing.

It is an uncommonly good example of the H-plan house, its aisled and originally open hall being of the same period of building as both cross wings. The whole frame, drawn as though restored insofar as would be necessary, is illustrated in Fig. 123. The jetties, built only on the front of the two wings, are soundly framed in the manner that in fact predominated during the later medieval period, which suggests an early advance in, and consolidation of, this feature. The concomitant first floors, however, show no such assurance and are mounted upon samson posts set into the planes of the cross walls. By this means the floors are supported quite independently of the walls, which with their storey posts support only the roofs. All wind bracing in the walls' planes was originally fitted in pairs insofar as the front elevation was concerned, which clearly indicates conscious decorative technique since single braces were considered adequate for the other walls. The crown post with its water-holding base is shown as Fig. 335, and a diagram of the service-wing hall, with both samson- and storey posts, as Fig. 124. In both cross wings the jointing of the common joists into the bridging-joists by means of bare-faced soffit tenons is of great importance. These were the only floors at the time of building and they assist with the dating by comparison with the choir stalls of Winchester Cathedral.

Fig. 124. Baythorne Hall, outer wall frame of a cross wing.

The Spire Scaffold, Salisbury Cathedral, Wiltshire

At the time of writing the dates of the upper parts of Salisbury tower and its spire are uncertain; Dr. Harvey has dated the great tower to *c.* 1334 (J. H. Harvey, 1961, 163). The designers responsible are equally uncertain, but both Elias of Dereham and Nicholas of Ely were involved from the first, according to the cited authority, and whether a master carpenter was also involved is a matter for speculation. Considered as a single work of masonry the tower and spire may well be the supreme exercise in Gothic attenuation, certainly in England, and possibly further afield.

Waiving all previous considerations concerning a period during which the spire was merely a stumpy lantern, and assuming that the building operation was carried through from base to cross without interruption, it is evident that timber scaffolding exceeding four hundred feet in height was virtually impossible, if only because of the great reduction in base area above the roofs. This fact led the unknown master concerned to scaffold his spire from within; the structure evidently rose by sections that overtopped the rising masonry cone (which contained about four thousand five hundred tons of Chilmark stone) and provided the workmen with octagonal stages of scaffold plank which still survive *in situ*. The upper six stages were supported by means of eight rib timbers, each designed to fit into the internal angles of the spire, doubtless with a view to damping wind pressure vibrations, which would be greatest towards the slender apex of the cone. The cap stone must obviously have been set from without, and external scaffolding must have been rigged from the internal structure for this purpose, afterwards being retrieved and lowered via the weather opening, which was placed at a height where a man's shoulder width prevents further internal ascent. A vertical section of the upper tower and spire, with its timber scaffold, is shown in Plate V.

The spire is 8ins. thick throughout, and its angles are contained in single pieces of Chilmark, which were necessarily of remarkable size, since they also had to furnish the projecting ornaments of the arrises. The great weights involved would appear to be more than the slender scaffolding could have supported, its own weight in oak and iron being estimated at about forty-five tons; but the materials were hoisted inside the tower's base area by the treadwheel windlass, which also survives in use.

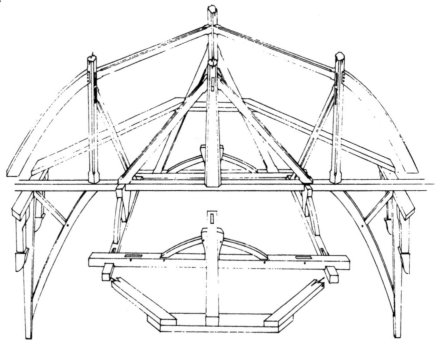

Fig. 125. Salisbury Cathedral, base of spire scaffold.

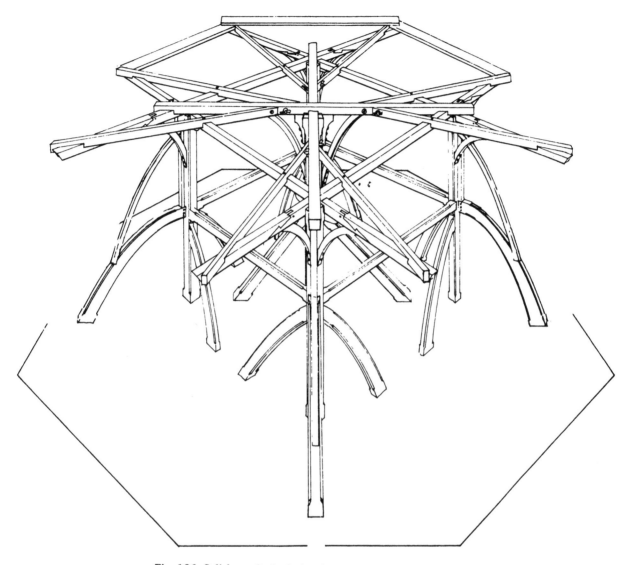

Fig. 126. Salisbury Cathedral, spire scaffold above parapet level.

Fig. 125 shows in perspective the manner in which the scaffold was mounted on top of over forty feet of simple timber tower, and was strangely devised to afford a 45deg. rotation of the four principal supports, curiously distributing its own weight upon both masonry walls and timber floor joists, wall pieces and corbels. Fig. 126 shows the elaborate basis of the next three stages of scaffolding, which stands on a floor at the level of the parapets. This structure, again, disperses its weight variously upon the one beneath and the external walls of the tower. The framing is amazingly complex and employs iron forelock bolts and spikes at numerous points; it supports, as shown, the first octagon of scaffold boards, the view from which must have been absolutely unnerving even for experienced masons. The next drawing (Fig. 127) shows the three subsequent stages, in order of ascent, the lower being the second above parapet level, and one of the most remarkable

Fig. 127 (*left*). Salisbury Cathedral, spire scaffold, the next three stages. Fig. 128 (*right*). Salisbury Cathedral, spire scaffold, remnant of lifting mechanism.

works of carpentry known. Finally, Fig 128 shows the remains of the original iron 'machine' by means of which the weight of the whole scaffold assembly (down to the point at the base of Fig. 127, where the central mast fits over the transoms with four toes) could be suspended from the capstone by using the iron cross, which penetrates the cap.

At the unknown date when this work was begun the master concerned introduced the scarf joint illustrated in Fig. 254. This seems to have been the first scarf ever to possess a bridled abutment; the reasons for its use at this point, or in this construction, are not yet determined, but its effect upon the subsequent development of scarfing was remarkable.

The Church of St. Mary the Virgin, Sheering, Essex

A church of undoubtedly Norman origins, possessing a remarkable and typologically rare roof of a design that must precede the gambrel roof of the 17th century; the two pitches in each of its slopes are of equal lengths. It is two bays long and has 15 common rafters in each bay, framed into the purlins which occupy the positions where the pitches change; these purlins are mounted on queen posts, and at the apex there is an elaborately sectioned ridge timber mounted on king posts. The tie beams are mounted on wall pieces which themselves stand on moulded timber corbels, and hanging knees are fitted into all structural angles (Fig. 129). The wall plates have the deeply hollow mouldings (Fig. 358) characteristic of the earlier 13th century, to which period this work is ascribed. The southern side purlin is scarfed with a splayed and tabled joint, placed immediately over the queen post.

Fig. 129. Sheering church, the nave roof.

The Roofs of the Nave Aisles, York Minster

The nave of York was built between 1291 and 1345, under the direction of Simon Mason (J. H. Harvey, 1961, 169), and the aisle roofs were possibly begun in *c.* 1300 and completed by *c.* 1320. These frames are low-pitched tie beam examples, so prolifically braced as to produce absolutely inflexible triangles, and the use of developed jowls indicates the relative lateness of their date. The scarf employed was still splayed with under-squinted butts, confirming the suggested final date for this work. The form of sole piece used for the common rafters is good, and so designed as to hook over the upper internal arris of the outer masonry wall (Fig. 130).

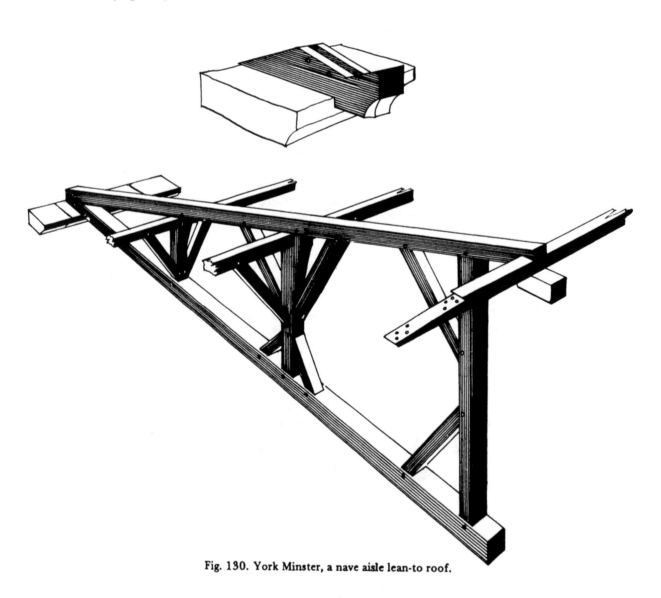

Fig. 130. York Minster, a nave aisle lean-to roof.

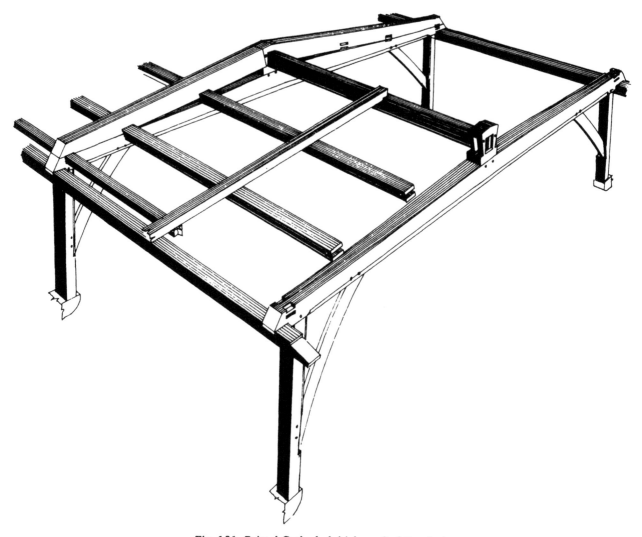

Fig. 131. Bristol Cathedral, high-roof of the choir.

The Choir Roof, Bristol Cathedral

The choir itself dates from between c. 1311 and c. 1340, and what survives of an ancient high roof above it is assumed to be the original. Fig. 131 shows one bay's length. The pitch is very low and the assemblies of tie beam, dwarf king post and principal-rafter virtually constitute 'built' camber beams, and presage the advocation of such roofs during the Perpendicular period that was to follow. The butted side purlins are so jointed as to enter both rafters and tie beams, thereby mechanically preventing any bending of either timber. The king posts have sunken faces, mortised for the massive ride pieces, and also bearings beneath their side mortises for the principals; six common rafters were fitted into each bay. The architect is not known.

147

The Church of St. Mary of Ottery, Devon

A justifiably well-known church having architectural affinities with Exeter Cathedral that have given rise to some conjecture as to the respective dates of the two designs, both of which have towers in transeptal positions. Adequate historical information seems to be available in the guide-book (J. A. Whitham, 1956, 8–9), according to which John de Grandisson was enthroned as Bishop of Exeter in 1328, and thereafter negotiated tirelessly with Rouen for the foundation of a collegiate church at Ottery St. Mary. Royal Licence was given for this foundation by Edward III on 15 December 1337, the college staff comprising 40 members. Prior to this a church was recorded as having been consecrated in 1259, to 'Sancte Marie Otery' in the register of Walter de Bronniscombe, Bishop of Exeter, 1257–80. The fabric of the collegiate church was modelled (according to Whitham) 'with much care upon Exeter Cathedral. It was completed about 1342'. The lower parts (two-thirds) of the transeptal towers are ascribed to the Early English period, and Bishop Bronniscombe, and the nave to the Decorated period, and Bishop Grandisson.

The little-known and hitherto unpublished timber high-roof covering the nave and linking the transeptal towers is of considerable interest in that it closely resembles the most eastern high-roof surviving in the cathedral. This latter roof, covering the four eastern bays of the cathedral, was roofed before 1300 (J. H. Harvey, 1974, 25), a fact that lends confirmation to the dating of the St. Mary's roof between 1300 and before the *c.* 1342 completion date for the fabric (sufficiently long to allow for the completion of the vaults beneath its covering), possibly *c.* 1330. The roof is a scissor-braced design chase-tenoned together, with its main run aligned from east to west and the two transeptal ridges valley-boarded against it. The frames are alternately collared with notch-lapped purlin struts and uncollared with raking purlin struts, the latter being the essential design of the eastern couples above the Exeter presbytery, begun by 1288, and the former resembling the later couples completed by *c.* 1310 as far as one bay west of the crossing. A single, flat wall plate was provided for the ashlar pieces while the rafters' feet were set upon suitably angled courses of ashlar. An example of each framed couple is illustrated in Fig. 132, in which the profile is of the notched and lap-jointed purlin struts placed centrally.

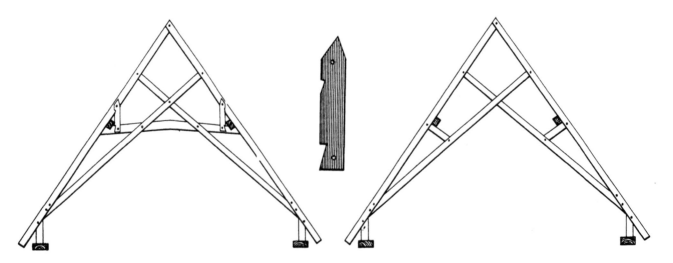

Fig. 132. Ottery St. Mary church, high-roof of the nave.

The Church of St. Peter, Goldhanger, Essex

A church with 11th-century origins and a nave roof in the Decorated style, which is in four bays, and has elaborate crown posts, tie beams of composite construction with deeply hollow mouldings, and wall pieces with pierced tracery spandrels. Most remarkable is the stop chamfering of every constituent timber in the roof, including the smallest. The roof is represented by one crown-post truss in Fig. 133.

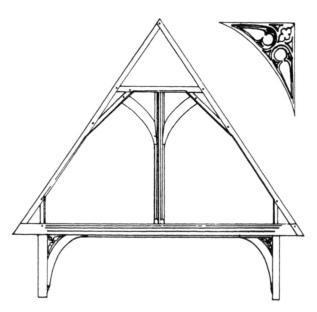

Fig. 133. Goldhanger church, section of nave roof.

Exeter Cathedral, Devon

According to the records (J. H. Harvey, 1961, 133; C. A. Hewett, 1974, 14) one bay of the nave was completed, west of the crossing, by 1310. 'There is clear structural evidence that the Norman nave was much lower than the present one, hence the old roof (of *c.* 1140–90) must have been taken down, and if any of it was re-used, it would have had to be a re-framing at the new higher wall-top level. As the master throughout this period (certainly to 1340) was Thomas Witney, one of the few proven examples of an architect who was both mason and carpenter, and an outstandingly advanced designer, one would not really expect any re-use, but rather a brand new roof designed *c.* 1325. To sum up: the only timber roofs made at Exeter Cathedral for the high main-spans would belong to the period: 1288–1342. The nave (except east bay), designed *c.* 1325 and probably all in position by 1342 (canvas was bought to make a temporary blocking of the great west window; this makes sense only if the timber roof was already up; the vaulting in stone is later, as usual). There is one absolutely precise date: the wooden vaults in the transepts (towers) were made in 1317.' (J. H. Harvey quoted in Hewett, 1974, 25.)

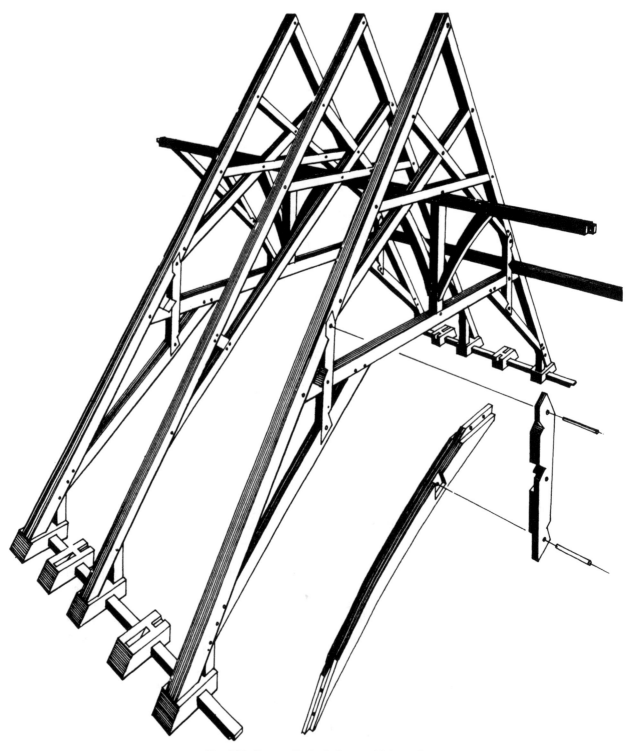

Fig. 134. Exeter Cathedral, nave high-roof.

The roof designed by Master Witney (Fig. 134) incorporates short, butted and tenoned collar purlins; the whole structure was obviously prefabricated and subsequently reared in bays of three common couples, when each bay was locked and stabilised by the rearing of a further crown-posted principal-couple. This roof extends along the great length of the nave and continues the side purlins of the earlier eastern high-roof, adding the crown posts (which are clearly designed to resist racking) and their central purlin. A feature peculiar to both eastern and western roofs is the method of wall plating, which was effected by short lengths— virtually distance pieces—against which the couples were reared. The roofs connecting the transeptal towers are probably a little later and, as illustrated in Fig. 135, they included a novel form of lap joint which may be defined as a 'counter-sallied cross halving', since that is what it is.

Fig. 135. Exeter Cathedral, transeptal roof.

The mock vaults which form ceilings to the transeptal towers are shown in Fig. 136, which represents the southern example. The ribs of these were suspended by vertical timbers from the elaborate platform assembly forming the floor above: the crown of the vault rises through the square void framed into this floor. The profiles of the rib timbers cannot be determined with certainty, in view of their height above the floor.

An important development in scarfing exists in the roof leading into the southern transeptal tower. This technique introduced the use of both iron spikes and forelock bolts into the stop-splayed scarf, and was used to produce rafters of the required lengths.

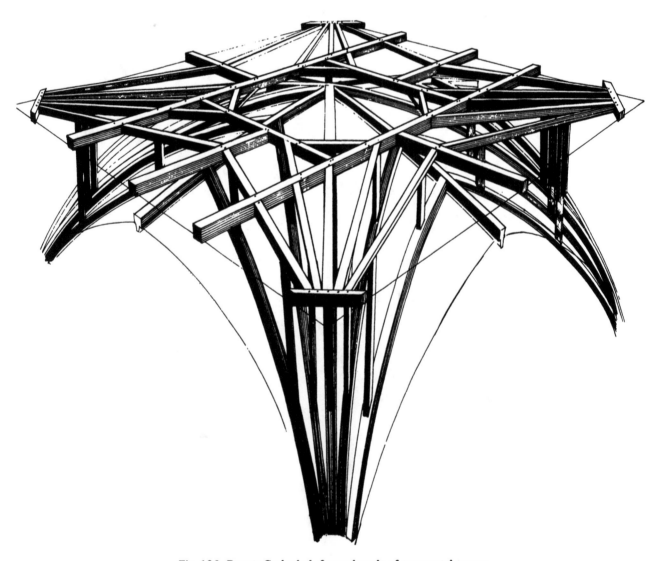

Fig. 136. Exeter Cathedral, framed vault of transeptal tower.

The retro-quire has close structural and other links with the lady chapel, which may have been completed by 1319, but was certainly so by 1326 (L. S. Colchester, letter to author, 1971). No named carpenter is known. This remarkable roof is illustrated in Fig. 137. It is hipped, and designed to encircle the west end of the lady chapel, which in plan forms an irregular octagon; further to confuse the situation it had to be mounted on the crowns of a bewildering complexity of vaults. The chosen solution to this problem was to mount 15 well-jowled posts at any points capable of supporting them, on either vaults' crowns or pockets, some of which spring as many as six arched braces from the same level. However, when viewed, this politic solution to a complex problem gives a strong sense of order and sound construction.

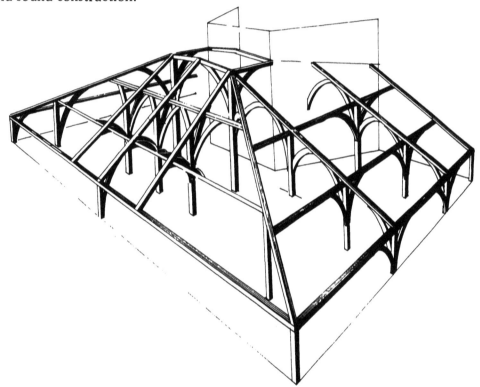

Fig. 137. Wells retro-quire, high-roof.

The lady chapel, which is vaulted and has a high-roof, presents another spectacular essay in timber roof framing (Fig. 138). The carpenter responsible wall-plated the irregularly octagonal plan, as illustrated, by means of a complex that incorporated wall pieces, two tie beams and numerous curved braces in the horizontal plane: from the two tie beams he tied what may be regarded as the ends of this plane figure, of necessity using down-curved ties, three to each end. King posts (one on each beam) support the ridge piece, and there are in addition almost all the structural features of many types of roof design, for it is an eclectic work.

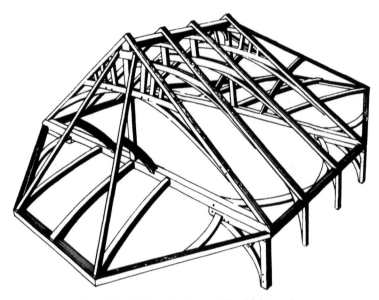

Fig. 138. Wells lady chapel, the high-roof.

The High-Roof over the Great South Transept, Lincoln Cathedral

The roofs of Lincoln are in a coherent series, and that over the southern transept may have derived from previous designs here. The building date for the substructure is known to be *c.* 1200-20 (J. H. Harvey, 1961, 141), but the identity of the master responsible is uncertain. Another firm date exists for the completion of the south transept gable: *c.* 1335-40. In this roof the tie beams were fitted to every fourth couple, and the whole structure is mounted upon double wall plates, the internal pair of which were given hewn face fillets in order to locate the sole pieces. This last was an excellent technique, but one that was rarely used and very expensive in terms of both timber and labour. Each couple was fitted with three collars, and the rafters were duplicated to the height of the uppermost one, with compressive struts in the two lower spaces. This design closely resembles that of the midstrey roof of the Abbot's barn at Glastonbury, which is not dated, but which may for this reason be roughly contemporary. A cross section of the Lincoln example is shown as Fig. 139.

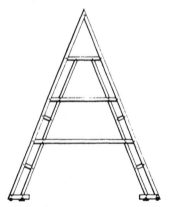

Fig. 139. Lincoln Cathedral,
south transept high-roof.

154

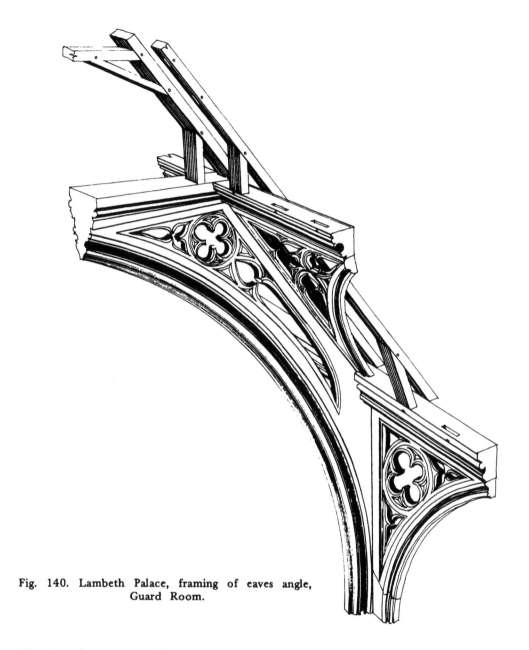

Fig. 140. Lambeth Palace, framing of eaves angle, Guard Room.

The Guard Room, Lambeth Palace, London

It is not known when the great hall at Lambeth was first built, but the beginning of the 14th century is considered a probable date (D. Mills, 1956, 22). A first-floor hall of this date survives and is known as the Guard Room, wherein, it is believed, the Archbishop housed such men-at-arms as he retained; the quality of the hall is, however, too high for it to have originated from such a function. It is built in four bays with five principal-couples, arched to their collars, with fine pierced oaken tracery beneath both side purlins and eaves plates (Fig. 140). The span of the roof is 27ft. and its bays are 14ft. in length. This is the best work of carpentry in Lambeth Palace, and can have few rivals anywhere.

155

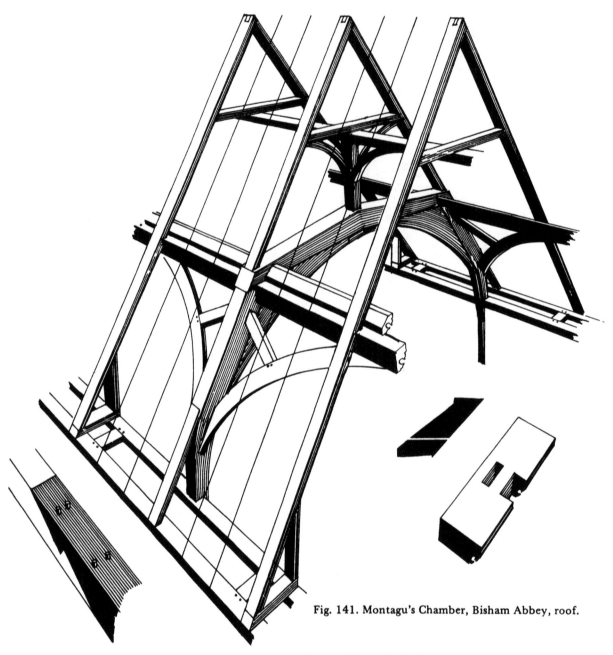

Fig. 141. Montagu's Chamber, Bisham Abbey, roof.

Roof to Montagu's Great Chamber, Bisham Abbey, Berkshire

The building of this range, which incorporates a cloister (*V.C.H.*, 1923, 139) and is stone-built, is known to have taken place after 1336, when William, Lord Montagu (1301–44) obtained possession of the manor. The mouldings worked on the side purlins confirm this dating of the work and are shown in Fig. 141, which also includes the scarf joint used. The roof consists of five bays and is without base ties, having elbowed eaves blades secured with knees to the half-height tie beams, on which a crown post and collar purlin assembly is mounted. The internal span is 23ft.

156

The High-Roof of the Nave, Winchester Cathedral, Hampshire

The dating of the nave is confusing because restyling was undertaken by Bishops William Edington from 1346–66, and William of Wykeham from 1367–1404; the architects were William Wyford from 1360–1403, and Robert Hull from 1400–02 (J. H. Harvey, 1961, 165). In addition the roof has suffered at least two major revisions, the former possibly relating to Wykeham's activities, and the latter introducing a different wall plating system that may relate to the restoration of 1905–12. Various features are left as evidence of this succession of events, including the end of at least one 'stone-fast' and circularly-sectioned tie beam, which possibly survived from the original Norman nave that was restyled. Technological type-series, however, suggest that the earliest surviving roof design is probably the one completed before the west front of the cathedral in *c.* 1360. This is illustrated in Fig. 142, which also shows the later queen post system that was introduced to avert racking; the scarf joint used for the wall plates is of the unusual type previously noted at Salisbury, which suggests a date in close chronological succession.

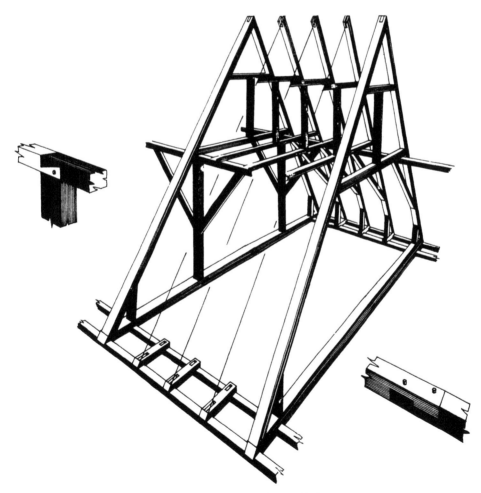

Fig. 142. Winchester Cathedral, high-roof of nave, scarfs enlarged.

157

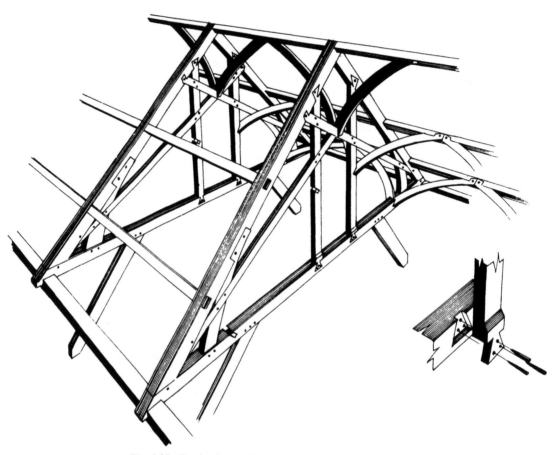

Fig. 143. Tewkesbury Abbey, high-roof, north transept.

The High-Roof of the North Transept, Tewkesbury Abbey, Gloucestershire

A remarkably elaborate and eclectic roof system survives here, and its survival testifies to its efficiency; the dating of the substructure does not assist with that of the carpentry, but technological details are comparable with the roof of the Wells retro-quire of *c.* 1329–45. Part of this is illustrated as Fig. 143, and details of the joints relating to the Wells roof as Fig. 144. Three bays exist, each having seven couples within a distance of 13ft. The tie beams are straight, and suspended at their centres by queen posts which have lap-dovetailed ends at both top and bottom. These queen posts are scissored, as shown in the drawing, and compassed struts resembling ashlar pieces are lap-jointed into the assembly. A ridge piece is socketed into the principal-rafters' apexes, and likewise two side purlins into each slope. The ridge and purlins are elaborately braced with curved timbers that spring from appropriate, but most unusual, points in the assembly—those to the ridge, for example, are stepped in bird's-mouth joints worked at the crossings of the scissor braces. Others, as the drawing shows, are stepped in similar sockets cut into either the queen posts or the tie beams.

158

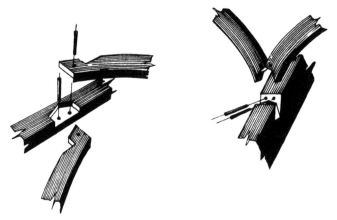

Fig. 144. Tewkesbury Abbey, high-roof, joints used.

Kennington's, Aveley, Essex

According to Morant 'The part of Keliton, formerly holden by Suein, is a large farm, belonging to the Lord Dacre, known at present by the name of Kennington. The possessors of it after Suein are unknown, till the reign of King Edward the Third, when it was in the Gernet family, owners of Wenyngton maner . . . Sir Henry Gernet, who died in the year 1345, held, jointly with his wife Joanna, one messuage, 120 acres of arable, 15 acres of meadow, and 6s. 8d. rent, in Alvethele, of the Prior of Prittlewell, by the service of 3s. *per annum*'. (P. Morant, 1816, 83.)

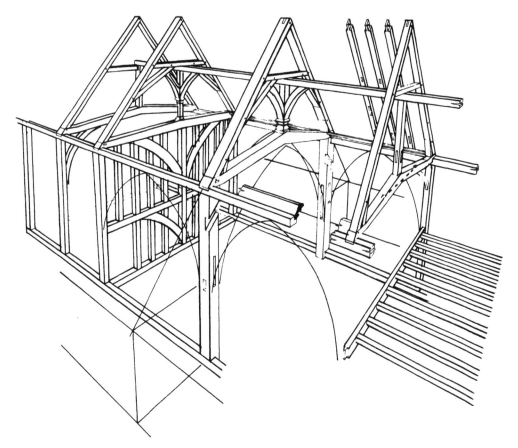

Fig. 145. Kennington's, perspective of surviving frame.

The surviving house has evidently been reduced in both length and width, but it was originally an aisled timber-framed hall of much style and pretension, designed during the period when scroll mouldings (which are its main ornament) were most popular, that is between 1260 and 1340 (H. Forrester, 1972, 31). It was rebuilt without its aisles at some date after *c.* 1375, when the newly-introduced edge-halved and bridled scarfs were used for its top plates. As originally framed, the solar end had a first floor in which the common joists were tenoned nearer to their soffits than to their centres, as was noted in connection with Stanton's Farmhouse (p. 136). The crown posts are illustrated as Figs. 331 and 332, and the entire frame as it survives as Fig. 145, which shows the jointing of the solar floor and the scarf used to extend the top plates.

Fig. 146. Ely Cathedral, angle post of lantern.

160

Fig. 147. Ely Cathedral, assembly of inner octagon sill.

The Octagon and Lantern, Ely Cathedral, Cambridgeshire

Polygonal ground plans were a feature peculiar to English cathedrals, and were particularly developed in respect of chapter houses. The roofing of these provoked some masterly works of carpentry, as has been noted at York; but for sheer size of undertaking the octagonal lantern of Ely is, without question, the supreme achievement of the English medieval period in polygonally-planned timber construction.

This huge and spectacular work had 'no direct parentage, and no immediate progeny' (J. H. Harvey, 1961, 82), and the inspiration and conception of this idea is generally attributed to the sacrist of the time, Alan of Walsingham. It is an unfortunate fact that the sacrist's rolls give 'no hints of any scheme or plan on which the building was proceding . . .' (F. R. Chapman, 1907, 14), and that the rolls for many years are missing. The building operation lasted from 1328 until 1342, and it seems from the rolls that Master William of Hurle ('Majister Carpentarius') was retained at the great cost of £8 per annum, 'worth something like £1,000 in the purchasing power of 1960' (J. H. Harvey, 1961, 82). He 'ordained' the work at the outset and held 'definite and continuous authority in the building'

161

(F. R. Chapman, 1907, 45). In addition Master John of Ramsey, a member of a family of Norwich masons, was Walsingham's mason, and a study of the fabric shows clearly that these two were working to a very detailed design (doubtless accurately drawn) from, at the latest, the time when the eight arches of the crossing were to be laid on their centring. This is evident from the placing of stone sill hooks in the masonry spandrels of those arches, at half their height, and thereafter from the accurate vertical trenches built into the masonry to hold the original posts of the eight angles (all of these are now removed).

Some general dimensions of the work are relevant; the internal diameter of the lantern is 30ft., the circuit of its sides 160ft., and its corner posts rise 50ft. above the roof. The central boss of the highest internal 'vault' is 152ft. 6ins. above the floor. It is definitely recorded that the carpenters took over the great scaffold, from which the masons had built the octagonal carcase; and that, according to Chapman, a 'great crane for lifting heavy weights' was in position when the timber structure was set up, and had been built for the purpose.

The present text is not a suitable place in which to detail the assembly methods used for this unique work, but a brief synopsis of it is given because a lack of methodical studies has shrouded the feat in mystery until the present time. The first timbers to be positioned were of necessity the eight great posts set within the internal angles of the octagonal carcase, and these were probably retained

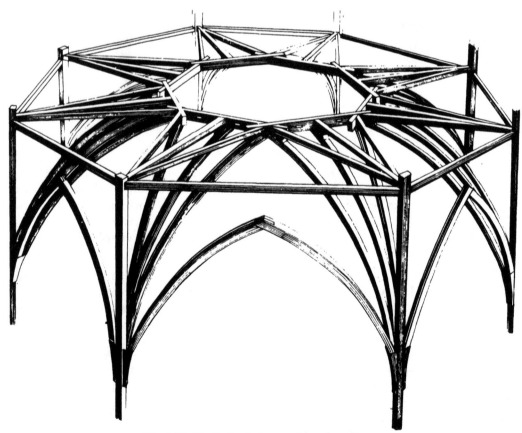

Fig. 148. Ely Cathedral, assembly of vault.

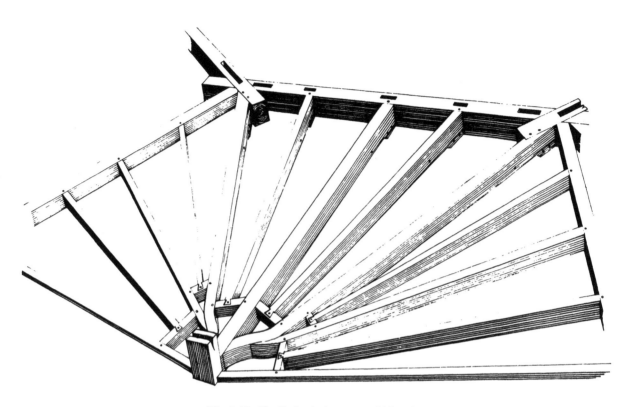

Fig. 149. Ely Cathedral, sector of 'floor'.

by timber tie-backs set into the sill hooks in the spandrels (Fig. 146). The straining timbers at higher levels were probably fitted by chase mortises to facilitate assembly; but since all of this original framing was removed during the 18th century, this must be an hypothesis. When the eight angle posts were in place, tied back and strained apart, the mounting of the internal octagonal void could be undertaken. For this purpose the 'great scaffold' the masons had passed on to the carpenters was used to the full, and the central space was bridged with fir trunks to provide a working platform at the base height of the lantern.

The chosen framing method prevented the assembly of corbelled triangular units, because their inner ends could not be stabilised until all eight were in place; and, as shown in Fig. 147, these were mounted as two successive operations, four units to each one. Complete stability was then assured by fitting the diagonal ribs of the vault (Fig. 148), when the completion of the vaulting with a tierceron rib to each joist of the platform ensued. A diagrammatic drawing of one-eighth of the joisting is shown in Fig. 149, in which it may be seen that each joist is so aligned as to support one vault rib. The plan in Fig. 150 shows the three points at which the huge posts set in the lantern's angles were split. This enabled the structure to be assembled first as half an octagonal tower, against its previously set raking shores, and then as two quarter-octagonal facets. This method enabled the great posts to be hoisted vertically into their positions and fitted and pegged, without their ever being inclined to any unmanageable extent during the operation.

163

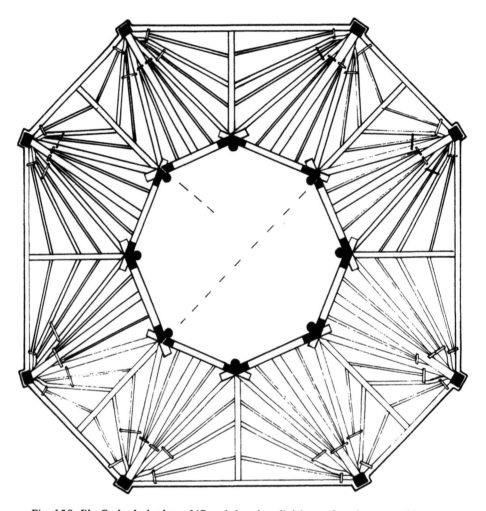

Fig. 150. Ely Cathedral, plan of 'floor' showing divisions of angle posts of lantern

The 'Stables', Church Farm, Fressingfield, Suffolk

A tenuous association with Ely may be suggested by the name of one of the priors there, John of Fressingfield, who died in 1338.

Part of a formerly open hall survives, comprising one gabled end, a spere-type transverse frame and one and a half bays of the hall range. At some unknown date this has been laterally divided by a clamp-mounted and lodged first floor. This is the most unnecessarily elaborate and decadent piece of visually spectacular carpentry known to the writer. The scarf joint used is an innovation of questionable merit, if any, employing a free tenon between square butts. The wind bracing is magnificent and must represent the limit to which this structural device was developed; it is not merely doubled in each panel area, but paired in depth, one brace being surfaced on both the inner and outer faces of the walls! In addition, the prodigious display of elegantly curved braces in the gable is disposed in a fan-like manner. All these features are illustrated in Fig. 151. A selection of the rich variety of mouldings, which indicate a date of c. 1330, is shown in Fig. 336.

164

Fig. 151. Fressingfield, the 'stables' at Church Farm.

The Church of All Saints, Messing, Essex

This is illustrated in Fig. 152. The roof can be dated to between 1344 and 1362 by reference to the heraldic achievement carved upon one of its sole pieces. This portrays the arms of the Baynard family, who held the manor of Messing from as early as *c.* 1272. The escutcheon itself is of 14th-century shape; and, in addition to the charges, it bears a label of five points—the first of the heraldic marks of cadence, indicating that the bearer was an eldest son. The only Baynard likely to have been responsible for this roof would have been the third Thomas Baynard, who would have borne the label between 1344 and 1362; and it is within this period that the roof must have been completed. The moulded wall plate profile is therefore significant, because firmly dated. The frame designed to clear the chancel arch is a single hammer beam unit not unusual in that situation; but the remainder of the range has principals with a smoothly curved archivolt. This curvature was achieved by cutting into both the rafters and the highly cambered collars. The fashion pieces set before the ashlar pieces were also curved. The surviving common rafters (there was a 19th-century restoration) were framed into seven cants, as illustrated.

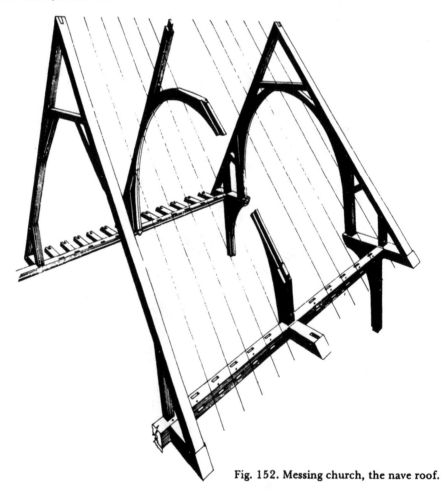

Fig. 152. Messing church, the nave roof.

166

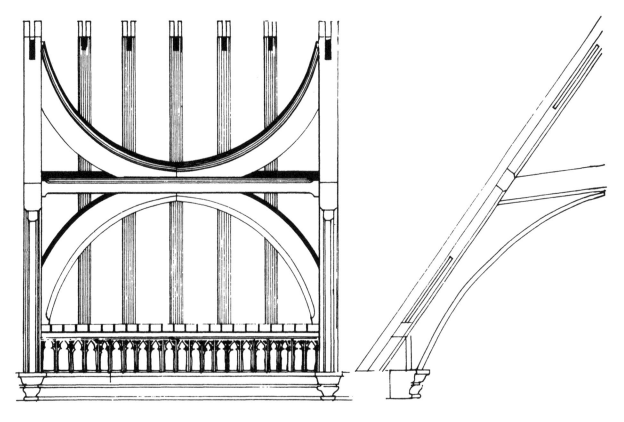

Fig. 153. Roof of the Chequer, Vicars' Close, Wells.

The Roof of the Chequer, Vicars' Close, Wells, Somerset

The dates of the various buildings of Vicars' Close are problematical, although three documents exist which suggest that a fair number were finished and in use by 1348. These documents are the will of Alice Swansee, dated 7 November 1348; a licence in mortmain, dated 3 December 1348, and a deed confirmed by the Dean and Chapter of Wells on 3 January 1348/9. This last refers specifically to 'the Vicars living in the new building erected by him and eating together in the common hall of the said building' which leaves the dates of the various other buildings, and of the vicars' houses, relatively uncertain. The Chequer leads north from the hall, at the same level, connecting it to the tower standing at the northern end which embodies parts (buttresses and arches at ground level) of an earlier tower of the Decorated period. Its roof is in four bays with braced collars and side purlins that have wind bracing beneath them and inverted wind bracing above them, as shown in Fig. 153. According to Mr. L. S. Colchester, formerly Cathedral Secretary of Wells, 'Favoured by a Wells carpenter between 1420-50, and almost certainly derived from the "inverted arches" under the Cathedral tower'.

167

The Roof of No. 22, Vicars' Close, Wells, Somerset

Although the dates of the vicars' houses are uncertain, as Mr. L. S. Colchester states: 'there is no reason to suppose that they were not completed in Bishop Ralph's lifetime'. In his will he leaves corn and cattle 'to the vicars of Wells dwelling in the houses built by me'. (Will dated 12 May 1363 at Lambeth [244 Islip] in S.R.S. xix, 1903, 286.) The design of the roof to No. 22 is common to all the houses, having as it does principal rafter-couples with arch-tied collars, and three purlins in each pitch, with a ridge piece. The side purlins are wind-braced in-pitch. The lowest of these is so set as to form an internal eaves cornice, and has what may be the earliest recorded example of the crenellated ornament (Fig. 154). All the purlins are fitted into the principal-rafters by means of tenons with soffit spurs, and in the case of the lowest ones this fitting was also scribed, as shown in the drawing. The arch-tying of these couples is also the earliest known example of that technique.

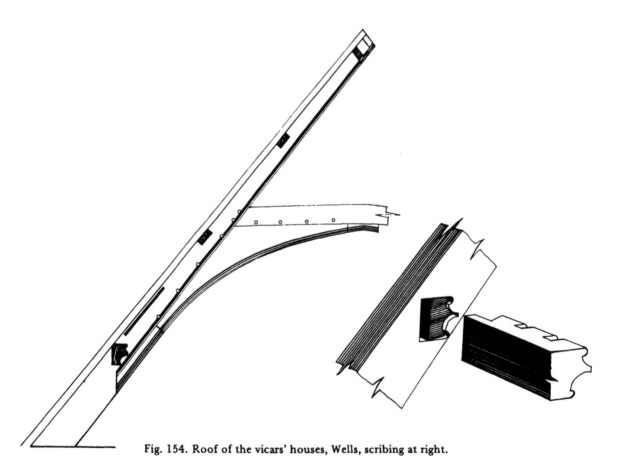

Fig. 154. Roof of the vicars' houses, Wells, scribing at right.

The Decorated Period—Summary

Two informed opinions of the Decorated period have been published in recent years, the first being to the effect that the 'architecture of England between 1250 and 1350 was, although the English do not know it, the most forward, the most important, and the most inspired in Europe' (Sir N. Pevsner, 1943, 128), and the second stating that 'the half century from 1285 to 1335 witnessed the most brilliant display of sheer inventiveness in the whole history of English medieval architecture' (Dr. P. Kidson, 1965, 106). It is unlikely to be coincidental that the carpenters' works of these years are all equally remarkable, and that of such historic carpentry as has survived for our appreciation, the most spectacular works date from then. It is even difficult to realise that the quoted authorities were not referring to carpentry, for example roofs, but were commenting on the styling and structural design of architectural carcases, which were invariably of stone.

Altogether almost fifty works have been described, all of which were designed and constructed within this period. They cover a wide range of types, a recent arrival among which seems to be the carpentered vault. The latter is a mere simulation, in timber, of a stone-mason's technique of making a stone ceiling for his stone-surfaced internal spaces. The motivation to develop such vaults as masonry produced (which probably culminated in that over Henry VII's chapel at Westminster), must have been the natural desire to produce interiors that seemed to be the work of masons alone. Such naves as those at Peterborough and Waltham Abbey are very difficult and less satisfying, for the eye follows the arcading upwards to meet with a transverse discontinuity of both material and structural method. The vaulted nave of Exeter, which is also of the Decorated period, is perhaps the antithesis of this, for the entire surface is uniform and every stone compressed. However, without carpenters both the construction and preservation of such vaults would have been impossible; they could only be laid and set upon wooden centring, and protected by timber high-roofs. It was mainly during this period that vaults were envisaged over spans that were impracticable, such as that of the nave of York Minster, or the octagon of Ely Cathedral; in these cases the carpenters provided timber-framed replicas of masons' vaulting because these were lighter, and their weight could be distributed in a variety of ways, unlike stone vaults. In other examples, such as that above the presbytery of St. Albans, lower costs seem to have been the main reason for choosing timber mock vaults instead of real ones. Whatever the reason for the choice, such framed replicas were in questionable taste; but, having said that, at their best—as in the Ely octagon—they are impressive. The St. Albans example is the earliest which survives and may be the first of its kind, and it is regrettable that its restorers did not see fit to retain every feature of its original design. It was evidently integral to its timber high-roof, and at two points along their curvature its ribs were suspended from wall posts, but little more can now be deduced; the rib sections, however, conform to the style of the period and are illustrated in Fig. 355.

Closely associated with such timber vaults are the polygonal chapter houses and their polygonal timber roofs or spires. It is in this context that a more general acceptance of curved timber for use in bracing is apparent, for it was forced upon any carpenter required to produce rib vaults; and during the middle years of the period Stifford's spire was designed with a rich display of interleaved curved braces that were never to be seen. Since they were polygonally planned, the Salisbury spire scaffold and the Ely lantern come within the same category; and considering their great size and uniqueness both must be assessed, alongside Herland's roof to Westminster Hall, as being among the supreme achievements of the English Middle Ages, if not also in the wider European context. It seems that in the spire scaffold and close to its base, the splayed-and-tabled scarf with a bridled abutment was first used, for some reason that cannot yet be deduced, and this may have led to the transition from splayed to edge-halved scarfs, as will be illustrated. Another principle applied to scarfing at Salisbury during this period was that of tonguing-and-grooving used in the top plates of the Old Deanery, which may indicate the work of one carpenter who was more influential than any other in that field. It was undoubtedly these innovations that led to the perfection of scarfing during the reign of Edward I.

Notched-lap-joints continued to be employed, but their use declined, and they were not always of the most efficient category. Yet among the final examples of their use, in the transeptal roofs at Exeter, another new development occurred—the counter-sallied cross halving. A single mainstream of developments can be traced in the field of high-roofing the main spans of great churches during this period, but several parallel courses also existed. Some such roof designs that can be firmly placed in the scheme, mainly sited in the south-east, appear to relate to French designs of roughly comparable times, but European studies would be necessary to interpret this fact. Among such designs are those of Chichester's high-roof, that over Salisbury's north porch, and the open roof surviving at St. Augustine's Abbey in Canterbury, to name but three. The dating of the Chichester example must depend on the dating of the earliest jowls with pronounced and splayed profiles, which occurred between c. 1260 and c. 1280 in England, but comparable information for the French examples is not readily available.

Concern with rafter racking was apparent in a majority of the roofs described, and numerous different systems were introduced to combat the failing: rafter bracing, for example, in the transepts of Westminster Abbey; collar purlins at Exeter, introduced by c. 1325 in the nave; and three such collar purlins above the Angel Choir at Lincoln. These various exercises at the highest level of building probably had an almost concurrent effect on the carpenters of parish churches, and one such result seems to have been the roof of the nave of White Roding church. The short bay lengths of cathedral roofing persisted during the early years of the period, and then attained their most refined form, that of branching tie beams seen at Salisbury; this same principle was also applied at Chichester to compressive struts.

It is a curious fact that once such structural features were perfected they were relegated to sporadic use and then seemingly forgotten, as was true of the notched lap, the branching tie or strut, and the perfect scarf that enjoyed a brief and expensive exploitation in these years. The eclectic craftsman had, by this time, a wide range of well-tried roofing systems available for permutation, and good examples exist in the roofs of Tewkesbury Abbey's north transept and the retro-quire of Wells Cathedral. Towards the end of the period roofs as innovative as that at Sheering church were appearing provincially, and equally without known precedent was the roof design above Bristol Cathedral, in which the principle of the built camber beam was implicit. An uninterrupted exploitation of grown eaves-angle timbers is illustrated by the roof over Montagu's Chamber at Bisham, where crown post, collar purlin and short bay length were all combined with eaves blades.

A variety of residential buildings with great and aisled halls represent the overall date range of the Decorated period, and despite their structural differences those enriched with mouldings conform to its style. Typologically arranged, the Bury at Clavering comes first, and thereafter both queen-posted designs, as at Merton's *aula custodis* and Gatehouse Farm, occur among a majority of crown-posted types. This wide structural variety is a reflection of those views quoted at the outset, and it is suggested that the works of masonry of this period have not more to offer than those of carpentry.

The development of aisled halls involved two important factors; the creation of storeyed end bays, for solar and service purposes, and the subsequent evolution of the jetty and the H-plan. Very little has ever been published concerning the development of timber first (or higher) floors or of their medieval concomitants, jetties. The earliest timber floors described, such as that in the church of St. Etheldreda, are lodged—that is, laid into their position and rested upon either stone outsets designed to receive them or the side girts of timber buildings. Such floors were commonly supported upon samson posts, placed under them to ensure rigidity, but the need for these cannot be established now except in cases where the joists were butted together over the posts, as at St. Etheldreda's. In many instances their existence indicates experimentation with an unfamiliar construction. This fact is important because the lodging of floor joists must have suggested the device of the jetty, or overhung upper storey; laying a set of joists across a pair of lower walls, whether of masonry or timber framing, probably familiarised structural carpenters with the fact that such projecting ends could easily support a super-structure. Because the majority of such floors (among the small total known) were laid transversely, in order that the joists required were not longer than the commonly available timber, it is to be expected that long-wall jetties appeared first, but no example is known that proves the supposition.

The most experimental example illustrated is the service wing at Tiptoft's Hall, and that was partly lodged and transverse, with longitudinal joists tenoned into the cross joist nearest the front wall; the necessary projection of the side girt was merely a short timber tenoned onto the storey post, which extended to the full height. No ground-storey wall was built under this jetty, but a samson post was placed at the centre of it instead. One cannot deduce a date for this work, but it must have occurred during the early Decorated period.

The next development, the ambitious provision of jetties to both ends of both 'wings' at Priory Place, was probably associated with the evolution of the H-plan. These floors were, again, primarily lodged, in this case for a third of their area, and had longitudinal joists tenoned into the two outer cross joists, producing the jetty projection. Experimentation with an unfamiliar mechanism is again evident, because in this house there were ground-storey walls framed beneath the jetties, but the projecting joists did not rest on their top timbers. Instead, they were alternately supported by rising braces from the studs or by their floor boards from above;

furthermore, it is evident that the amount of projection achieved was minimal, and the quantity of timber used greater than was afterwards found necessary.

In Baythorne Hall the type of jetty framing generally seen had been evolved with the system of jointing illustrated in Fig. 292, but the building is very important for two other reasons, both associated with the flooring. The first of these is the use of soffit tenons for the projecting joists (the type of joint used for the stalls of Winchester Cathedral from 1307-9), and the second, and more important, is the use of samson posts to carry the floor; these were not placed centrally as in earlier cases, such as Chesterford Manor House of *c.* 1190, but in the line of the wall. As a result, this house was designed with storey posts to carry the upper storey and roof, and samson posts to carry the first floor, indicating doubt on the part of the carpenter as to the ability of the storey posts to support both. It must therefore date closely to the Winchester stalls, and later than either Tiptoft's Hall or Priory Place, both of which should fit chronologically between Little Chesterford, with its purely lodged and posted floor, and Baythorne Hall.

The most archaic building among the selected examples to have a fully integral first floor is the solar wing of Wynter's Armourie (Fig. 98), wherein the binding joist was tenoned into the storey posts in the manner that was to become normal, and only the bracing of the frame and spacing of its studs betray the date of the design. Such early storey posts were, however, excessively weakened by the triple through-mortising necessary at floor level.

A development suggested by several of the examples cited is the elevation of the hall to first-floor level, frequently associated with queen-posted roof designs, as at Merton's *aula custodis*, the Fressingfield 'stables', Gatehouse Farmhouse, and the priory building (not mentioned previously) at Campsea Ash in Suffolk. It has been thought that this device was aimed at clearing the ground floors of free-standing posts. Such designs depended on the availability of very long binding joists. Fully 'compassed' roof designs, such as that over the Lincoln Greyfriars' church and the nave of Messing church, were numerous in the period; no explanation is offered for them and, as with the halls, it is considered that none is needed, since their appearance evidently pleased their patrons and carpenters alike.

Chapter Six

Examples from the Perpendicular Period (*c*. 1350 to *c*. 1450)

St. Clere's Hall, St. Osyth's, Essex

This a large, isolated house on a moated site south-east of the Priory and not far from St. Osyth's Creek. According to Morant, 'it appears that these lands, called St. Clere's-wic and St. Clere's-park, were held under the Earls of Oxford, . . . by Philip de St. Osyth, in 1273. Next William St. Clere, and his son and heir, John, in 1334'. This John possibly built the existing open and aisled timber hall, which is still open to its ridge and has two cross wings; the service wing is arguably contemporary with the hall, but the solar is obviously later, and is possibly associated with

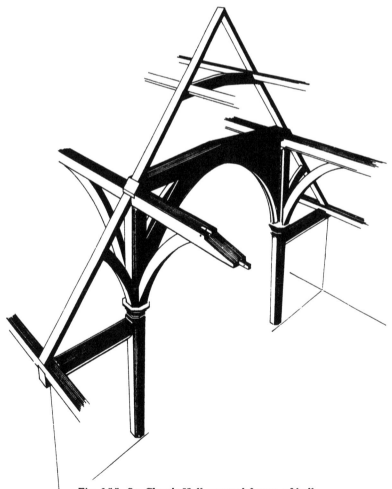

Fig. 155. St. Clere's Hall, central frame of hall.

'Thomas Darcy, of Tolleshunt Tregoos Esq., who died 27th. October 1558', holding two messuages (P. Morant, 1816, 459). The central frame spanning the hall is shown as Fig. 155, and the service wing as Fig. 156, the latter illustrating the conjunction of the two contemporary ranges. A timber sample from one of the transverse arch braces was carbon-dated by Professor R. Berger to a probable date of A.D. 1350 $\pm$60 years. The hall roof has side purlins, the eaves are of the reversed-assembly type with supine lap dovetails, and the scarfing throughout is splayed with bridled abutments—in this case both square and vertical, with four face pegs and two edge pegs. A large upper chamber in the service wing was also open to the ridge, as shown in Fig. 156, and spanned by an interesting collar-arched tie beam (Fig. 157). This is probably the earliest known example of such a tie, which constitutes a pair of tying-rafters and acts in the manner of a fixed caliper across the span.

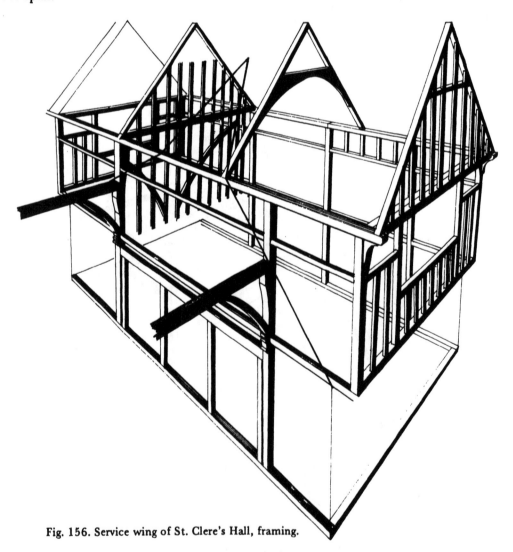

Fig. 156. Service wing of St. Clere's Hall, framing.

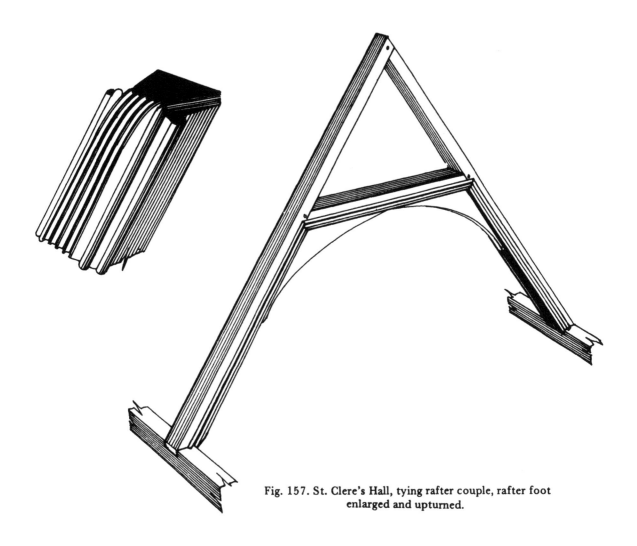

Fig. 157. St. Clere's Hall, tying rafter couple, rafter foot
enlarged and upturned.

The Barn, Prior's Hall, Widdington, Essex

One bay of this barn is illustrated in Plate XI. The building is eight bays long, and this necessitated frequent scarfing of the top plates, which was effected by the joint shown as Fig. 260. This is probably earlier than a similar one used at St. Clere's Hall, and an earlier date is proposed for the barn because the scarf is keyed. The spacing of the rafters was so calculated as evenly to divide the barn's length, and as a result the couples are not in any fixed relationship with the tie beams, to whose flanks they are rebated to varying depths. As a result, no rafters can be defined as being either common or principal-rafters. The outshut ties were fitted with hanging knees in their eaves angles, which were of the reversed-assembly type; this is an early example of their use. The jowls in this frame are elegant and of pronounced profiles. The fitting of the outshut purlins is of interest, for they were mounted on struts from the shores behind the posts, indicating an attempt to divide the roof's weight between the main posts and the ground sills of the outer walls, which were mounted on flint plinths.

175

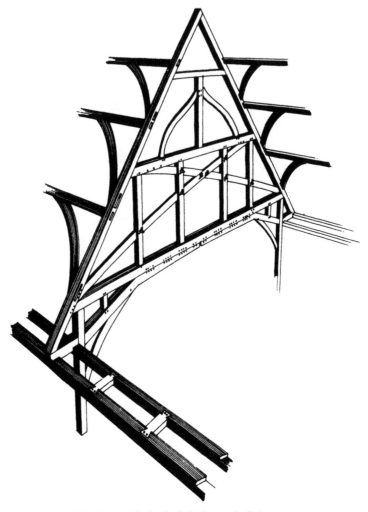

Fig. 158. Winchester Cathedral, high-roof of the presbytery.

The High-Roof of the Presbytery, Winchester Cathedral, Hampshire

A roof of complex design, as shown in Fig. 158, concerning the date of which there has been some disagreement; the substructure, however, is dated between 1315 and 1360, and its architect was Thomas of Witney (J. H. Harvey, 1961, 165). To this Thomas Berty added a clerestory, between *c.* 1520 and 1532; the high-roof is considered to date from the earlier building operation. No structural connection was seen between this roof and the framed vaulting under it, but it was evidently designed to cover vaults because its wall posts are alternately long and short, to fit either the crowns or the pockets. Structurally this was a strange undertaking, using great quantities of large, cranked oaks, possibly because that was what was readily available. The tie beams were of the built category and were unusually rigid, being secured along their entire length by profusely pegged free tenons. The posts mounted on the ties were each of two timbers, between which the scissor braces were trapped, and the whole forelock-bolted; it was then topped by a mountant, without any lengthwide connections. Three side purlins, with curving bracing, were fitted in each pitch.

176

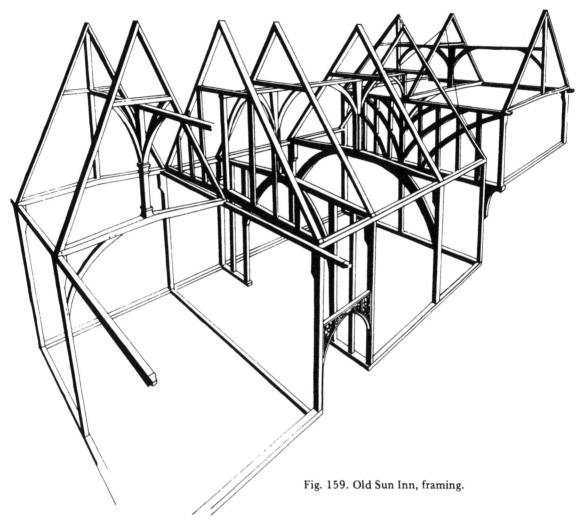

Fig. 159. Old Sun Inn, framing.

The Old Sun Inn, Saffron Walden, Essex

This is no longer an inn, but it retains its former name. It is H-plan, has jettied cross˙wings, and appears to be of one construction, the framing of which is shown as Fig. 159. The hall was never aisled and was fitted with a spere frame separating the cross entry from the hall; one entry door case has survived and is illustrated in Fig. 160, which also shows the pierced tracery of the service-wing lights. Together these features suggest a date of *c.* 1350 to 1360 (H. Forrester, letter to author).

Fig. 160. Old Sun Inn, doorway and window details.

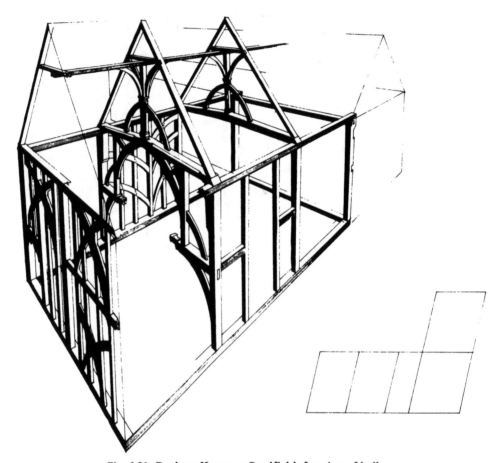

Fig. 161. Durham House at Bardfield, framing of hall.

Durham House, Great Bardfield, Essex

This is a building sited at the corner where a minor road meets the High Street; the importance of this junction is its obtuse angularity—an angularity that was carefully followed by the architect. This is illustrated in Fig. 161 on the plan to the right of the framing. The timbers were worked to the same out-of-square sections, and this even applied to the crown post which is shown in Fig. 343. The hall was initially open, and was fitted with windows of full height. A decorative but non-functional replica of hammerbeams was built into it, the ends of the hammers being carved with what seem to be portraits of the original builder and his wife (the latter is shown in Fig. 162). The wall framing of the hall is remarkable in that it has both triple- and saltire bracing, which suggests a decadent eclecticism as early as the later 14th century. The crown-post mouldings do not assist with the task of dating, but neither do they conflict with the date of the style of the head-dress, and a date *c.* 1370 is proposed for the building.

Fig. 162. Durham House, portrait
on hammerbeam end.

Bridge House, Fyfield, Essex

This is a small tenement in a loosely integrated village, which at the time of initial occupation possibly combined both trade and agriculture. It is an example of a small open hall which may represent a type common during the 14th century. The frame, insofar as it has survived, is shown in Fig. 163. The manner of fitting both collar and braces into the same housing with nail-headed pegs is illustrated in Fig. 341, together with the well-cut details of the crown post.

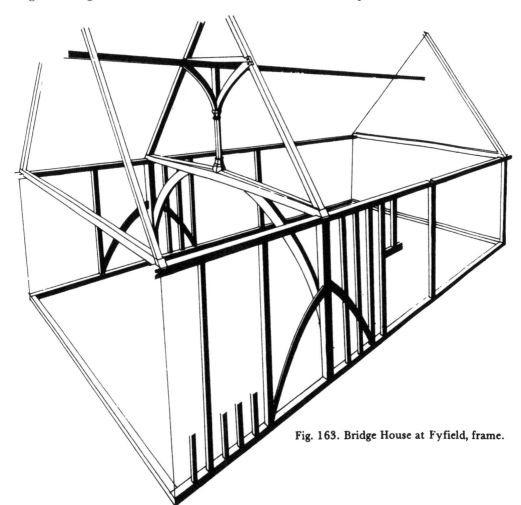

Fig. 163. Bridge House at Fyfield, frame.

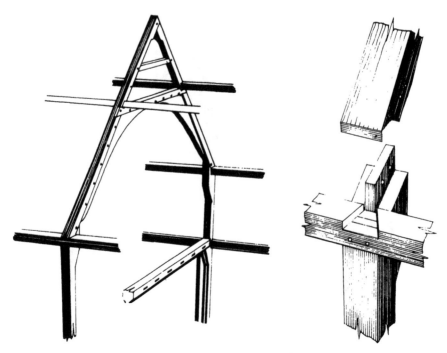

Fig. 164. Rayne Hall, tying rafters.

Rayne Hall, Rayne, Essex

This is a large H-plan house, the uniformity of which proves upon examination to be the result of numerous successive constructions, all with the H-plan in view. The manor 'was the part which was holden by Hugh de Montfort. He was the youngest son of Turstin de Bastenburc, a Norman, commonly called Hugh with the Beard . . . Thomas de Welles, held the same estate, and died in 1315. Walter Welles until 1325. Joan, daughter of Walter Welles, brought it to William de Rushbrooke, or Royssbrooke. He was living in 1362' (P. Morant, 1816, 400). In 1486 the estate passed into the possession of Sir William Capel. The original hall range may be attributed to either Walter Welles or to William de Rushbrooke, since it is clearly 14th-century; and in view of its floor-joist joints (which are barefaced soffit tenons) the earlier of the two seems more probable.

The length, in bays, of the hall, which was always on the ground floor, and originally front-wall jettied, with a correspondingly large upper chamber above it, cannot now be determined. Only three bays survive, one of which is terminal in position and narrow enough to imply a former chimney or smoke bay. The roof has side purlins and cranked braces, both across to its collars and in-pitch to the purlins, its greatest interest being in the direct conjunction of the storey-post jowls and the principal-rafters. Examination proves these to be another example of the semi-hexagonal tie in the form of the collar, its braces and the rafters and post heads. The relevant structural features of the main range are illustrated as Fig. 164.

The existing solar wing is evidently datable to Sir William Capel, who was also responsible for the handsome red-brick tower of the adjacent parish church.

180

The Roof over the Triforium to the North Transept, York Minster

This example (Fig. 165) is datable to between *c.* 1361 and 1373 (Dr. E. A. Gee, verbal to author). Technically all the rafters are common, since of uniform dimensions, and the roof closely resembles the earlier example at Lincoln, since the only concern was to maintain the flatness of the single pitch by direct transmission of the weight on to the masonry. The use of the jowled end for the raking strut places this work later than *c.* 1250, when the upstand appeared at Cressing. This fact also ensures that no date coeval with the completion of the transept carcase can be envisaged, since that was constructed *c.* 1230.

Fig. 165. York Minster, roof of north transept triforium.

Barnard's Inn, High Holborn, London

Here the open hall survives, concerning which Sir N. Pevsner states: 'It formed part of the London Mansion of John Mackworth, Dean of Lincoln who died in 1451. He left it to the Dean and Chapter who leased it to Lionel Barnard, who allowed students of Law to use it. It dates from the late fourteenth century' (Sir N. Pevsner, 1973, 335). The hall is of three bays and is framed in timber; it retains its timber louvred 'fumer' or smoke vent (L. Salzman, 1952, 465). The two transverse frames are apparently pairs of base crucks with tie beams supporting a crown-post and collar-purlin roof, of seven cants, with sloping ashlar pieces (Fig. 166). These crown posts are important, surviving as they do in the capital. They are illustrated in Fig. 347.

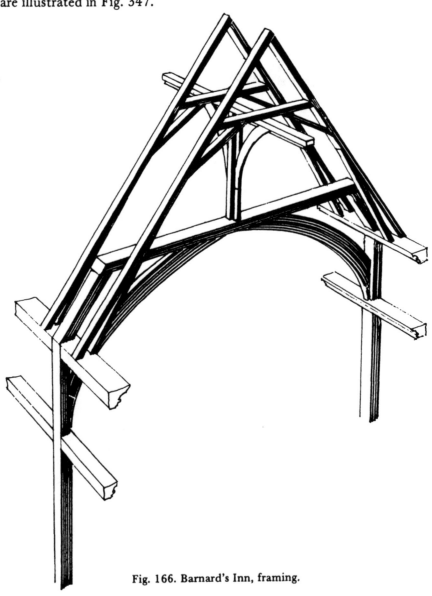

Fig. 166. Barnard's Inn, framing.

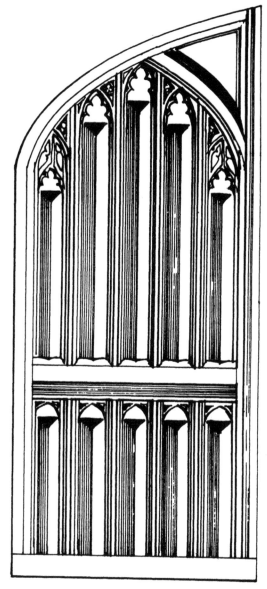

Fig. 167. Door leaf from Charterhouse Square.

The Charterhouse, London

This was one of the works of Henry Yevele, datable to 1371 (J. H. Harvey, 1944, 32). It was a Carthusian monastery designed round a 300ft. square, to which the existing square may relate. Little of Yevele's work remains above ground, as the cited authority states. The pair of oaken door leaves that today give access to the square are medieval (Fig. 167). The durns were cut from grown bends and together they form a single composition. Each panel is treated as a cinquefoil above the lock rail, despite the fact that hump-faced planks were used between the muntins, which necessitated chamfering of the planks' tops to provide a flat surface for the cusping. The bases of these planks were scribed to meet the upper edges of the rails. The panels of the lower series were treated as four-centred arches with sunken spandrels.

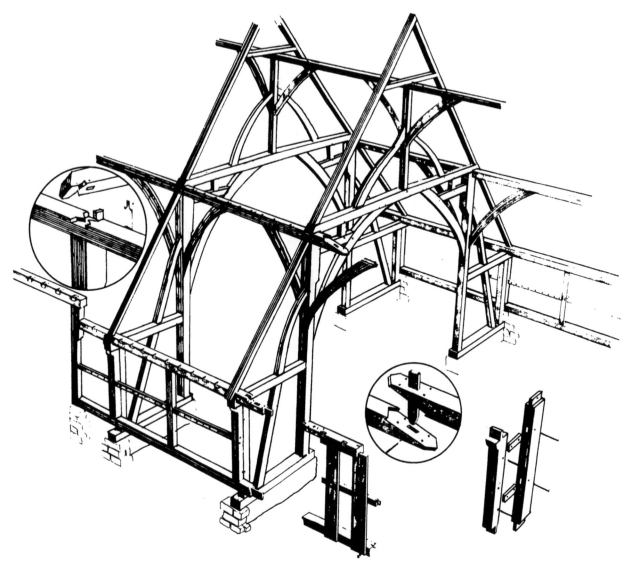

Fig. 168. Frindsbury, the barn.

The Barn, Manor Farm, Frindsbury, Kent

A structure of great length, being 210ft. from end to end and comprising 13 bays and additional *culatia,* or terminal outshuts. This is illustrated in Fig. 168, with some structural details. It originally had but one midstrey, crossing at the seventh bay. The site was formerly a demesne of St. Andrew of Rochester. The main-span assembly was normal and the eaves assembly reversed, the roof being crown-posted; all main posts were cut with pronounced but severely practical jowls and had their arch braces elaborately strutted, as shown in the drawing. The framing of the wall posts is also shown; these were in pairs, free-tenoned together. The weatherboarding largely survives and is vertically applied, as was the case of the similar barn at Upminster in Essex (C. A. Hewett, 1969, 123). The scarfing is of the stop-splayed variety with sallied butts and face keys; this joint is used in miniature, complete

184

with key, high up on the collar purlin, indicating a very high standard of jointing that was consistently maintained throughout the building operation, without undue regard to expense. The tie beams at main span were seated in the curious type of joint illustrated, which can be defined as a lapped and counter-notched crossing; this joint has also been noted in connection with the *c.* 1200 top plates of the barn at Paul's Hall in Essex. The date proposed by Mr. Rigold (S. E. Rigold, 1966, 11) was *c.* 1300 or earlier, and the radiocarbon date from Professors Horn and Berger was A.D. 1400 $\pm$ 60 (*Radiocarbon*, 1968, 410). Further research is desirable.

The Choir Roof, Carlisle Cathedral

One of the finest surviving specimens of the arched-to-collar type, Carlisle is also dated with some precision to the years 1363–95, during the episcopacy of Bishop Appleby and the life of the architect John Lewyn, although the latter is less certain (J. H. Harvey, 1961, 123). This roof always supported a timber ceiling, which survives, spanning a little over thirty feet, with the arms of the subscribers affixed to it. This is illustrated by a single couple in Fig. 169.

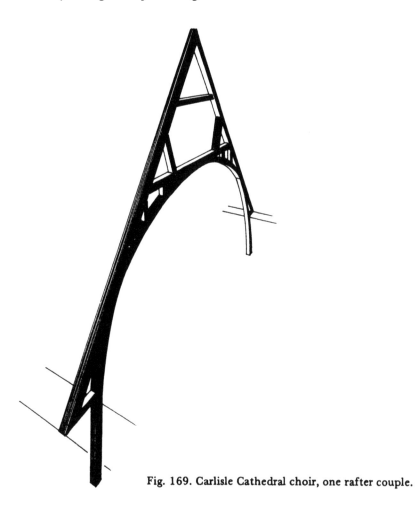

Fig. 169. Carlisle Cathedral choir, one rafter couple.

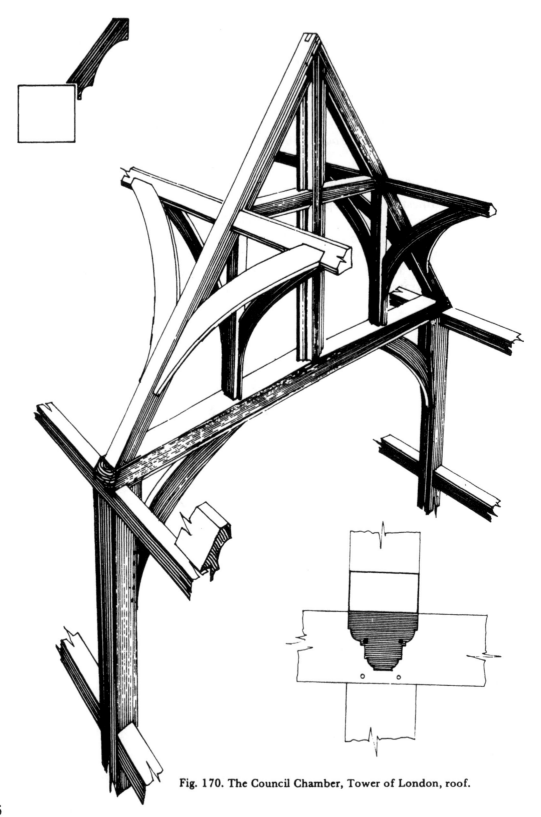

Fig. 170. The Council Chamber, Tower of London, roof.

186

The Council Chamber, The Queen's House, Tower of London

One wing of this L-plan building incorporates what is now known as the Council Chamber, which seems to have been of the great hall type originally, but was laterally divided by a first floor at some subsequent date. The external walls were framed heavily in open panels, the latter being broken only by ogee braces of little functional value which probably presage the fashion for serpentine braces that was to become widespread towards the end of the 16th century. Three bays of the roof survive in a mutilated condition, and a hypothetical reconstruction of it is shown as Fig. 170. The inset details are: top left, the cross section of the eaves cornice fitted internally; and lower right, the scribed housing for a former (but intruded) bridging joist. The jointing of the roof frames is highly complex, as are the cross sections of the timbers, and hypothetical internal economies for their assembly are shown as Figs. 171 and 172.

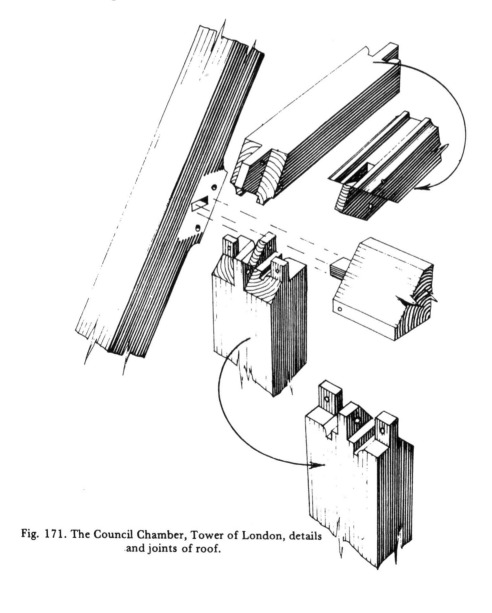

Fig. 171. The Council Chamber, Tower of London, details and joints of roof.

187

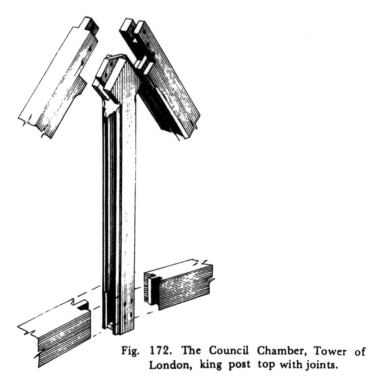

Fig. 172. The Council Chamber, Tower of London, king post top with joints.

The king posts were designed to act as pendants from the rafters' apexes and assist the queen posts in supporting the side purlins, the collars were designed to act in compression, and the whole lavishly braced. There has been some conjecture in the past as to whether this was originally a hammerbeam roof, but the joint analysis described disposes of the problem. The moulded profiles of these timbers are predominantly hollows, worked upon returned cross sections; and the sunken channel shown on the underside of the collars may here be appearing for the first time. A date between *c.* 1370, in view of the double hollows, and *c.* 1580 in respect of the complete absence of ovolos, is proposed.

The Roof, Westminster Hall, London

The supreme work of Master Hugh Herland, carpenter, dated 1394–1400 (J. H. Harvey, 1971, 104). This Dr. Harvey describes as 'the greatest single work of art of the whole of the European Middle Ages. No such comparable achievement in the fields of mechanics and aesthetics remains elsewhere, nor is there any evidence for such a feat having ever existed'. This roof is usually defined as belonging to the single hammerbeam category, but it is equally dependent upon its collar arch. In view of its enormous size, which Sir Banister Fletcher observed 'covers an area of nearly half an acre, and is one of the largest timber roofs, unsupported by pillars, in the world' (Sir Banister Fletcher, 1956, 451), a new structural principle was necessary. This was the use of composite timbers, or 'built' components, a type of carpentry that had always been needed for spans in excess of the longest available timbers, but which was never so fully exploited as at Westminster Hall. This is illustrated in Plates VI and VII.

188

North Triforium Roof, Norwich Cathedral, Norfolk

A lean-to roof that does not relate to others of its kind, it is a strange and highly ingenious structure that must date from between *c.* 1370 and 1510, since its tenons do not have diminished haunches; but no fabric dates assist with this matter. Dr. Harvey wrote: 'This certainly belongs to the raised triforium of the late fifteenth century, part of the masonry was in progress in 1472, so a date of *c.* 1475 is the likeliest' (D. J. Steward, 'Notes on Norwich Cathedral' in *Archaeol. Jnl.*, XXXII, 1875, p. 44). Part of the roof is illustrated in Fig. 173. Every rafter comprised three timbers, of which the first two formed a low-pitched ridge roof, and the third converted it into a lean-to roof. The internal economy of this triple joint has not been ascertained, and the roof, which was the last medieval one left at Norwich, has been destroyed since examination. No tying of the span was provided and the stability depended entirely upon the accuracy, and efficiency, of the triple central assembly. A ridge piece and two side purlins were fitted, the sections of which are illustrated, together with the wall-plate moulding.

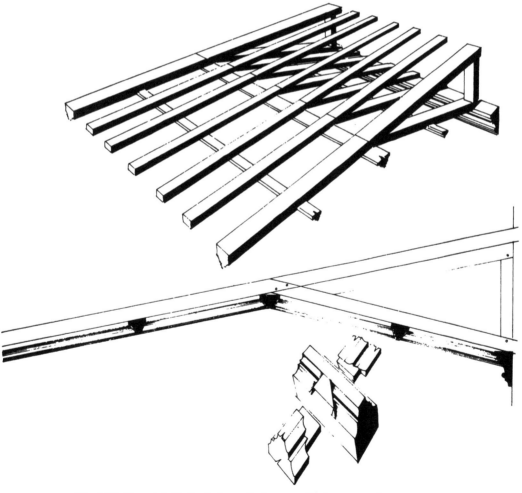

Fig. 173. Norwich Cathedral, roof of nave triforium, now demolished.

Roof of The Monks' Dormitory, Durham Cathedral

This is a low-pitched ridge roof that is firmly dated between 1398 and 1404 (The Ven. C. J. Stranks, 1971, 20), which provides evidence for the use of obtuse ridged roofs by an apparent minority of carpenters from early times. Very long timber had to be available for this, since it is a tie-beam roof with king posts, wall pieces and arch braces (Fig. 174). The boxed ties fitted to restrain the arch braces are a curious feature, decorated with carved fleurs-de-lis at their lower ends; the ridge piece is also ornate. The span is very wide and the bay lengths short, each bay having only five common rafters.

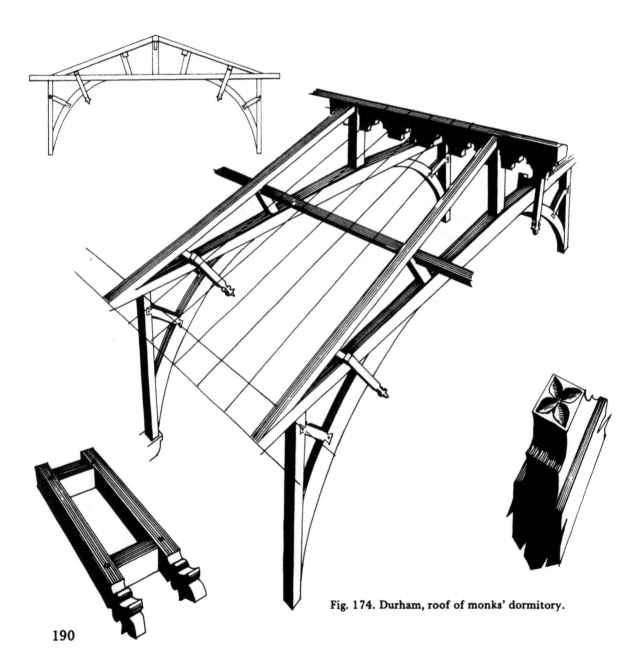

Fig. 174. Durham, roof of monks' dormitory.

190

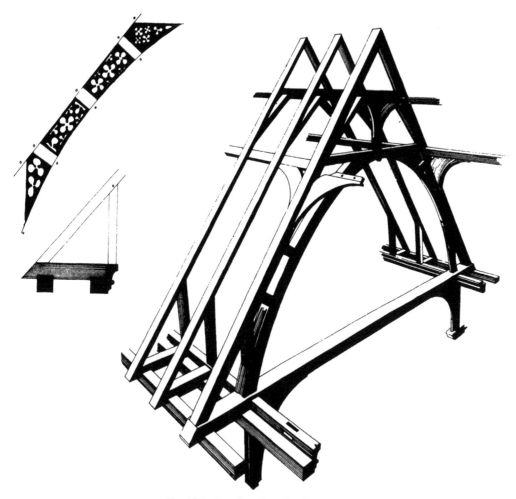

Fig. 175. Cressing church, the nave roof.

The Nave Roof, Church of All Saints, Cressing, Essex

This fabric was recently proved by excavation to have had origins long before the Conquest, and to have undergone numerous restylings. The nave roof, however, is among the most elaborate permutations of structural principles; it probably dates from the end of the 14th century, a period when many nave windows were also renewed. Part of it is illustrated in Fig. 175. The principles here combined were: tie beams on braced wall pieces, collar arches well strutted to their principals, high crown posts with collar purlin, and a main range of common rafters framed into seven cants—to which profusion side purlins with in-pitch bracing were added! Roofs such as this have considerable merit and they illustrate the last intelligent and objective designs that sought to continue the greatest and earliest developments exemplified by the cathedral high-roofs of the early 13th century. Only the pierced infill panels of tracery at Cressing leave something to be desired, possibly because no accomplished carver was available to do the work.

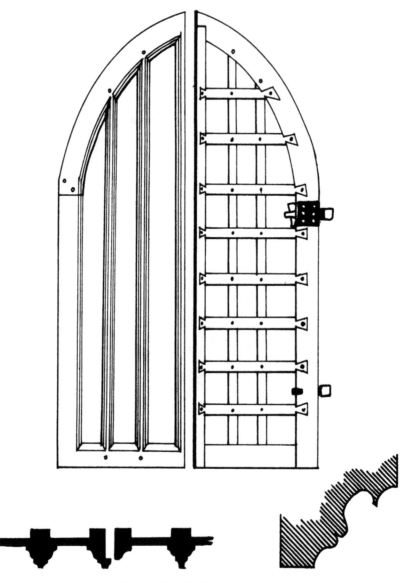

Fig. 176. Tolleshunt D'Arcy, door leaf from the church, with sections.

The South Door Leaves, Church of St. Nicholas, Tolleshunt D'Arcy, Essex

The outer and inner faces of the left-hand leaf are illustrated in Fig. 176. The last major styling of this church was Perpendicular before the restoration of 1897. The date of its south door and timber roof (which are of very good carpentry, without any of the decadent frivolities often found in this period) is *c.* 1450. They are hung from jointed durns and have two stiles and a rebated head timber, all with planed mouldings on the outer faces. The rear has eight fully dovetailed ledges and a sill, the latter being tenoned into both head- and harr durn.

192

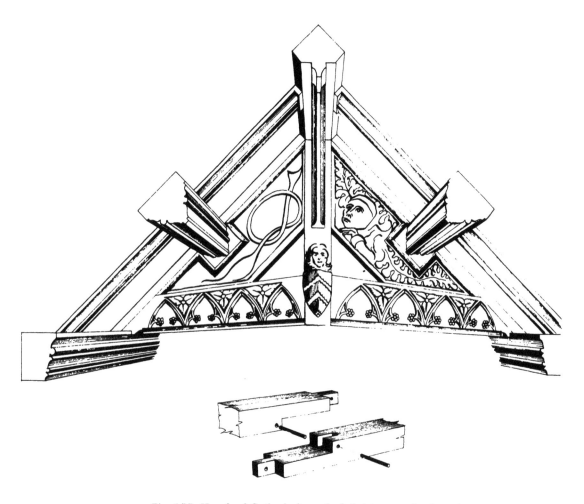

Fig. 177. Hereford Cathedral, roof of cloister, scarf enlarged.

The Cloister Roof, Hereford Cathedral

This is illustrated in Fig. 177. A king post assembly with remarkably deep principal-rafters, side purlins and ridge piece, the cross sections being well moulded (as is emphasised in the drawing); its precise date is not known, despite the heraldic carving upon the arcaded tie beam; however, it is known that the east walk of the cloisters was designed by Thomas Denyar and built between 1412 and 1418 (J. H. Harvey, 1961, 137). The scarf joint used for the wall plates is shown below the drawing, and is edge-halved and bridle-butted, as was almost ubiquitous during the Perpendicular period.

193

Tymperleys, Trinity Street, Colchester, Essex

Of this building Morant said: 'Dr. William Gilberd's own house in this (Holy Trinity) parish, anciently called Lanseles, and Tymperley's, or Tympernell's (Old Taxation) is the same as Serjeant Price, the late Recorder of this Borough lived in and now belongs to Thomas Clamtree, Esq. George Horseman and Frances, his wife, daughter and heir of Roger Tymperley, sold it in 1539 to Richard Weston, with a croft of an acre and a half, gardens, and three rentaries thereto adjoining, lying in Trinity and St. Mary's (Court Rolls, 31 Henry VIII., roll 14)' (P. Morant, 1768, 117). It is also known to have been the birthplace of Dr. Gilberd, physician to Elizabeth I, who was born in 1543 and was buried at Holy Trinity in 1603.

What survives today is a two-bay open and upper hall, with crown-post roof, and a fully-framed front-wall jetty; the floor joists have soffit spurs. The profiles are illustrated in Fig. 348, and the type of floor joint in Fig. 294.

The North Door Leaf, Church of St. Michael, Fobbing, Essex

This door is interesting mainly on account of its iron strap hinges, which are not proper subject matter for the present text; but they do illustrate the way in which the 12th-century 'anchor' type of ride strap developed into one of its subsequent forms, and this particular example must date from the 13th century, as indeed the Royal Commission suspected (R.C.H.M. 1921, Vol. IV, 45). The anomaly is the fitting of straight iron straps across such irregular timber surfaces, contact being achieved only at the high spots; for this reason it is considered that the wooden leaf must be of a later date than the irons. The planks used are the most complex noted among v-edged examples, and they were probably manufactured early in the 15th century. (Fig. 178.)

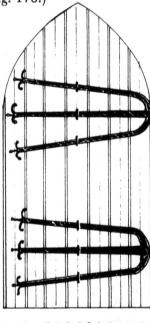

Fig. 178. Fobbing church, north door leaf.

The Beaufort Tower, The Hospital of St. Cross, Winchester, Hampshire

This hospital was founded by Bishop Henry de Blois in 1136 (P. Cave, 1970, 7), and claims to be Britain's oldest surviving charitable institution. It is still housed in a remarkably complete complex of medieval buildings, which includes the great church. The Beaufort Tower, associated with Cardinal Beaufort, who was Bishop of Winchester for 43 years, was built between 1404 and 1407, and is of considerable interest because the joists of its floors had tenons with housed soffit shoulders, the earliest of which are in the first floor and may date close to the commencement in 1404. (Fig. 295.)

The Library Roof, Wells Cathedral, Somerset

The history of the Wells library begins at least as early as 1298, when the term appears in an ordinance of the chapter; but thereafter its siting and the dating of its accommodation and equipment are confused. It is now housed above the east walk of the cloister, a position that is confirmed by the will of Bishop Bubwith, dated 11 October 1424, in which moneys were left for the building of a new library extending for the length of that arm of the cloister. Both the internal cloister wall and the existing roof are of two clearly distinguishable building operations, and the southern six bays of roofing have rafters with barefaced soffit tenons, whilst the northern eight bays have rafters with tenons mounted on spurred shoulders (Fig. 296). These rafters and their joints are reconcilable with the date of Bubwith's building, which is believed to have been completed by *c.* 1433 (L. S. Colchester, 1978, 15).

Chichele's Tower, Lambeth Palace, London

This tower, which has also been known as the Lollard's Tower, and the Water Tower, was built between 1434 and 1435, and detailed accounts concerning its costs have survived in Lambeth Court Roll No. 562. Its walls, which attain a height of 50 feet, were built of Kentish ragstone by Thomas atte Hille, and in 1434 16 bundles of straw were bought to protect them from the frosts; the total costs amounted to £192 19s. 4½d. (Sir N. Pevsner, 1973, 281). The first floor of the tower is quartered, on plan, and had four common joists to each quarter of area. Their cross sections are elaborate (as shown in Fig. 297), a fact that made their jointing difficult; here again we have an example of the conflict between the Perpendicular style and sound construction in timber. Scribing was the chosen jointing method for the shoulders, and the tenons were cut between the roll mouldings—the only part of the timber thick enough for that purpose.

The South Wing Roof, Horham Hall, Thaxted, Essex

Of this complex building Morant writes: 'this is supposed to be part of the two fees and a half which the heirs of Walter de Acre held in Thaxted, Chaure, and Brokesheued, under Richard de Clare, who died in 1262. It was afterwards in the considerable family surnamed de Wanton . . . William de Wanton, died 1347, held the maners of Chaureth and Horham' (P. Morant, 1816, 440). Another publication at the Essex Record Office (in Box T.I., and without references) states that Robert

Large, Lord Mayor of London, possessed the manor in 1439, and that the earliest surviving work is attributable either to Large or to his son Richard. Concerning a later period, during which the existing Great Hall was built, Sir. N. Pevsner records the fact that Sir John Cutte, Treasurer to the Household of Henry VIII, acquired the manor in 1502, and died in 1520 leaving the building unfinished. The date stone shows the year 1572, which probably marked the building's eventual completion—a date relevant to the roof over the Great Hall.

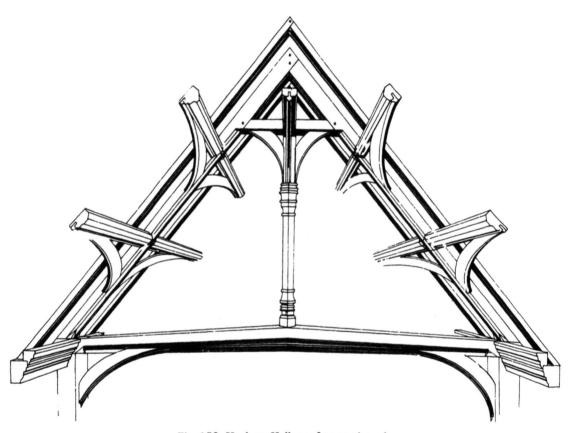

Fig. 179. Horham Hall, roof to service wing.

The surviving south wing, which obviously constitutes the remains of an earlier house than that of Sir John Cutte, probably represents the building of Robert Large. This was originally a wholly timber-framed wing of four bays' length, jettied to the east, and beautifully constructed with richly-moulded timbers. The central truss of its roof is illustrated as Fig. 179. It is unusual in that the crown post is diagonally set upon the tie beam, but more importantly it represents the fusion of two roof types; that with a central- or collar purlin, and that with side purlins. The mid 15th century is an appropriate period for such permutations of principles, and in view of the direct connection with London a date close to Large's acquisition is proposed for this work.

196

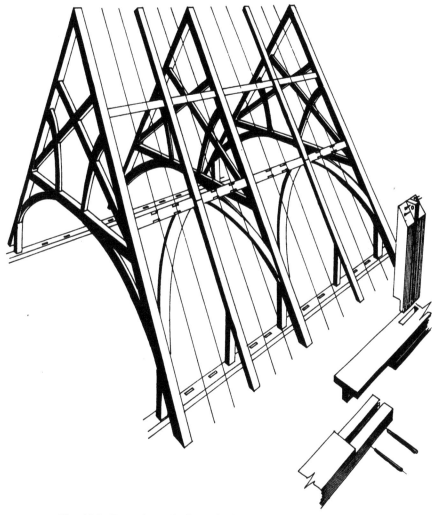

Fig. 180. Canterbury Cathedral, high-roof to north-west transept.

The High-Roof of the North-West Transept, Canterbury Cathedral, Kent

This is the sole surviving medieval main-span roof at Canterbury (Fig. 180). The transept beneath it was built by Richard Beke between the years 1448 and 1455 (J. H. Harvey, 1961, 121). This is a design which incorporates many earlier principles and which may justly be considered as one of the 15th century's permutation roofs; it embodies collar arches, scissor bracing that is itself scissored, and side purlins with in-pitch bracing. The form of scarf joint used (shown inset) is the same as was employed for the barns at Little Wymondley in Hertfordshire and Harmondsworth, Middlesex.

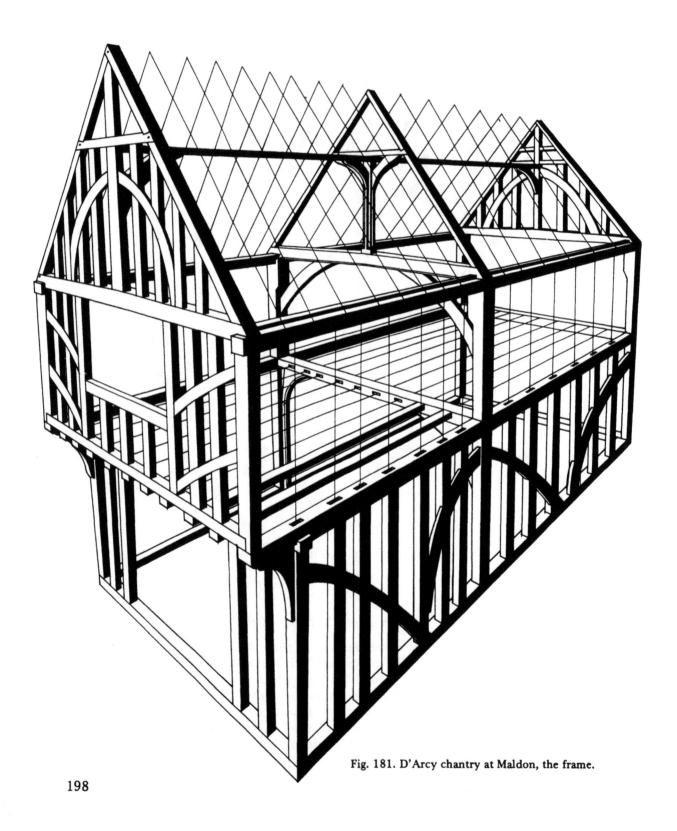

Fig. 181. D'Arcy chantry at Maldon, the frame.

198

The Chantry Priests' House, Maldon, Essex

I am indebted to Mr. M. C. Wadhams for the documentary research concerning this building (M. C. Wadhams, 1975, 213), and also for permission to reproduce his drawings of it. The vicarage of the parish of All Saints appears to be an H-plan house, but it is evidently the result of two building operations because the framing of the cross wings is different; the western one has paired external wind bracing on its first storey, the eastern has single braces. Examination confirmed the difference, and subsequently established that there were two builds, the first having been a small and jettied range with external stair, complete in itself, to which the hall range and solar cross wing were added. Since this is today a vicarage situated close to the church it was considered probable that so small a residence could only have been adequate for a celibate priest, and this was confirmed by Mr. Wadhams's researches.

Sir Robert D'Arcy died in 1448, having ordained that his executors should set up 'as quickly as possible' a chantry to be called 'Darcyeschaunterye' in the parish church of All Saints. This was to have two chaplains who were to celebrate the mass daily for the souls of D'Arcy and various other persons. The chaplains were to be provided with 'a Messuage and a garden and an acre of land in Maldon', and that messuage is today the western cross wing of the vicarage. The two-bay timber frame of this is illustrated in Fig. 181, and the several datable features in Fig. 182. Internal murals include stencilled slipped trefoils, and the sacred monogram *IHC*, the trefoils symbolising the Holy Trinity (to whom the chantry was dedicated). The floor joists were fitted by means of central tenons with housed soffit shoulders, and the edge-halved bridle-butted scarf for the purlin had splayed halvings; the roof was mounted upon a cross-quadrate crown post.

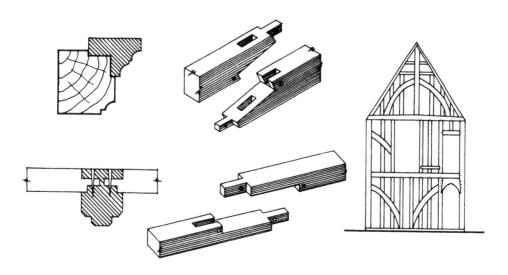

Fig. 182. D'Arcy chantry, structural details.

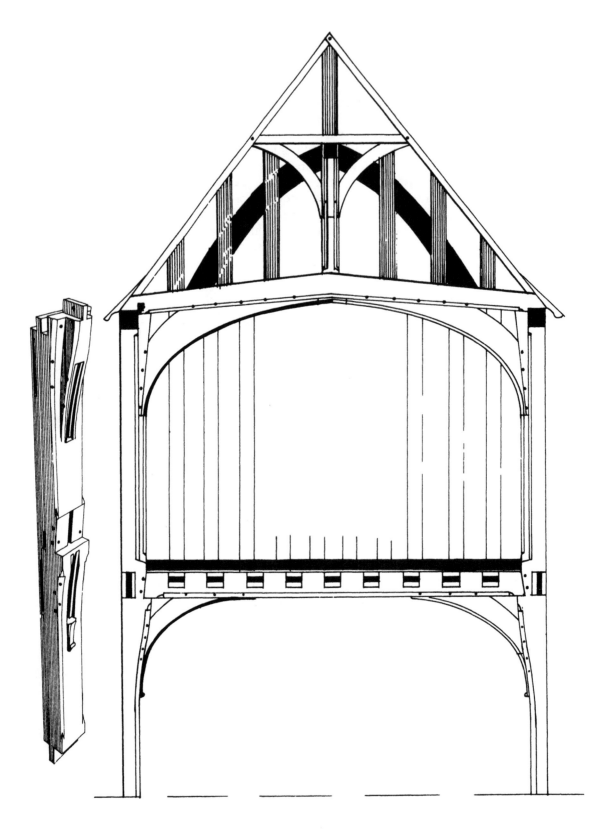

Fig. 183. Childe's chantry at Witham, section through priest's house; 15th-century storey post.

House Named 'Mole End', Chipping Hill, Witham, Essex

This timber building, of high quality and small size, is significant in the context of the development of the diminished haunch as an adjunct to the floor joists' end tenons; there are documentary grounds for its having been the residence provided for a chantry priest during the mid 15th century. Of this building Morant records: 'The other Chantry was called *St. Mary's Chantry*, alias Jennett Childes; and was endowed with a messuage and garden near the church-yard; an estate called Olivers, and divers lands, meadows, &c. in Witham, Wickham, Hatfield-Peverall, and Boreham; to find a priest daily to sing mass in this church, at the altar of our blessed Lady'. In the *Calendar of Patent Rolls* (Edward VI, Vol. II, 1548–1549, H.M.S.O., p. 84) reference is made to 'the messuage and garden in tenure of Edmund Stanbanke, clerk, in Wytham, next the churchyard there, all which belonged to Jennet Childes chantry'. As to the foundation of the chantry, it is stated (H. J. Rowles, B.A., *Essex Review*, Vol. 40, 1931, p. 12) that the 'Joan Child chantry dedicated to the Blessed Virgin Mary, was added to the south side in 1444', and a date between then and *c.* 1450 is proposed for the house. This was apparently of two bays, affording a chamber on both the ground and the first storey, and with a stone-lintelled fireplace of high quality flanking the western wall which was designed to heat the upper chamber. The latter was open to its apex, with a fine crown-post roof. The transverse section is illustrated in Fig. 183, and the joists' end joints in Fig. 298, these were central tenons, the face- and soffit shoulders of which formed a single spur—the direct forerunner of the diminished haunch.

The Belfry, Church of St. Lawrence, Blackmore, Essex

'Blackmore possesses one of the most impressive, if not the most impressive, of all the timber towers of England' (Sir. N. Pevsner, 1956, 76). This is illustrated in Plate IX, sectioned in both length and width to scale. The timbers used were enormous, and it is surprising that no documentation concerning the considerable felling and transportation of trees that must have been involved is known. The structure consists of three units, and these units are so designed that the second passes, covers and strengthens the mounting of the third, which emerges as the bell turret with its tall brooch spire. The cladding of the walls is of vertical boards applied to rails, as noted at the Frindsbury barn and elsewhere (C. A. Hewett, 1969, 123). The lap joint used for the tie beams was the same as that used at Cressing in 1250, indicating that some superlative joints were retained in use despite their cost. The priory, of which this church incorporates fragments, was dissolved in 1527; and the belfry seems to be a monastic work undertaken before the Dissolution, after which the church became parochial.

The Perpendicular Period—Summary

Among the 11 hall-type buildings described that were built during this period, St. Clere's Hall best illustrates the continuing and developing tradition of large rural manor houses; it survives sufficiently intact to establish that the H-plan design, with open hall and storeyed cross wings, was the prevailing type, and that the jettying and floor framing of such buildings had advanced technically. It also shows a predilection on the part of the patron for the latest fashion in decorative treatment of timber, and the adoption, by the carpenter concerned, of what must have been the most up-to-date jointing methods. Furthermore, the large chamber provided above the buttery and pantry in the service wing was built with a central tying rafter couple which is among the earliest known of its type.

The other buildings that are similar, Place House at Bardfield and the Old Sun Inn at Saffron Walden, serve to illustrate the variation in quality that was common to all periods. Both these buildings contain simulated details: in the roofs of the Sun Inn, for example, there are braces between crown post and purlin that are nailed in place, while the hammerbeams of Place House are purely decorative. Decadence had set in relatively early during the 14th century, possibly for the first time in English carpentry. The best example is the so-called 'stables' at Fressingfield where many of the timbers have no function, such as tie beams which possess no end joints. This phenomenon cannot be explained by a lack of funds or skilled labour, since both were evidently available; it was a matter of taste and aesthetic discrimination (or a lack of both). It seems that such deceptive works were effected during the closing years of the Decorated period, and, for what the surviving evidence is worth, not during preceding periods. It can, of course, be justifiably argued that more 14th-century works have survived than those of the previous century; but, nevertheless, no deceptive techniques dating from the Early English period have yet been recorded.

An example of these techniques which is common to both St. Clere's Hall and Fressingfield 'stables', is the fitting of planks into the spandrel voids of curved braces so as to create the appearance of solid hanging knees. This technique was so prevalent that it was applied to the small spandrels of some braces, high beneath the first floor, which was built into the late Saxon tower of Little Bardfield church at some unknown date during the 14th century.

The major houses usually had crown-posted roof designs, but two of them did not, and they are important for that reason; both St. Clere's Hall and Rayne Hall (the latter while in its original form) were fitted with side-purlin roofs, and both had tying rafter couples. Bridge House at Fyfield illustrates the way in which the great manor-houses were copied; they evidently set the pattern for people who built themselves houses during the period, and in most cases both the freedom and the necessary capital to do this probably indicate the growth of the yeoman class. This building had an open hall spanned by a transverse frame of considerable ostentation, of which the crown post is shown in Fig. 341; the evidence has largely been removed by alterations, but indicates the original provision of storeyed end bays, for solar and service purposes. Another category of small, but essentially high-quality houses was created by the endowment of chantries, such as that of Lord D'Arcy in Maldon (Fig. 181). These survive in some number, and were designed to give all the necessary accommodation that a celibate priest needed for comfort. As in the Maldon example, such buildings sometimes became enlarged by additions until the H-plan resulted, and a clue as to the origins of the smaller cross wing is often to be found in the traces of a former external stair to the first floor. Two other examples are the house called the Old Vicarage at Owls Hill in Terling, and Chantry House at Stebbing, both in Essex.

Among the recognisable developments which took place during the early 15th century was the combination of previously tested roof designs, to produce hybrids that were often of surprising complexity. Two examples of this are provided by what was once the solar wing of Horham Hall at Thaxted, and the roof over the nave of the church at Cressing (again, both in Essex). The former is important in that the two most distinct roof designs—crown posts with a collar purlin, and side purlins—were combined into one, a type of which at least two other examples are known in the barns of Rookwood Hall at Abbess Roding in Essex. The side-purlin design had very early origins, and it finally superseded the central- or collar-purlin designs completely. In the roof of Cressing church five separate principles were combined, and it seems that these hybrid roofs were among the better works of carpentry of the Perpendicular period, combining as they did several good designs from the past in structurally sound and aesthetically pleasing roofs. However, it is suggested that these did not represent the Perpendicular style, and were an aspect of the lingering transition from Curvilinear Decorated into Perpendicular—for which the appropriate roof was of the almost flat type called camber beam.

The camber beam is well represented by the various roofs of the great church of Saffron Walden (see Plate VIII), which were structurally little more than floors; but they were appropriate for the embattled and parapetted style which prevailed. It is of interest that the unknown master who designed the high-roof over the choir of Bristol Cathedral between c. 1311 and c. 1340 had produced what seem to be the best examples of camber beams which survive. These had all the extreme inflexibility that built, or composite, timbers can possess, while the majority of later roofs of this type relied on the availability of trees of uncommon lengths to bridge the spans. The roof of the Council Chamber in the Tower of London is an example, resulting from royal patronage, of a roof system combining several principles: king posts that were apex-suspended, and among the first known of that kind; and queen posts and side purlins with bracing in-pitch, and in the vertical plane. Also in London is *Barnard's* inn, which provides a valuable instance of the use of base crucks, in combination with a crown-post roof, and is apparently the sole survivor of that type in the capital.

The supreme work of carpentry for this and all periods was designed by Hugh Herland for Richard II; this was the new roof of Westminster Hall, completed in 1400, and illustrated in Plates VI and VII. This amazing masterpiece has been little published or investigated, being inaccessible due to its great height. Nothing is known concerning the tecnhiques by which its enormous 'built' components were assembled, and it may well be that so many far-reaching restorations have been undertaken that little information could now be recovered as to its construction and assembly methods. As has been stated, 'It is strange that so little public notice is taken of this roof; it is the most outstanding individual work in the whole history of English art, yet it ranks low in the list of London's historic attractions for the visitor. *Its true rank is with, or above, the Pyramids, as a Wonder of the World'* (J. H. Harvey, 1948, 24), author's italics. Elsewhere the same authority has said that 'Herland's oak roof remains as the greatest single work of art of the whole of the European Middle Ages. No such combined achievement in the fields of mechanics and aesthetics remains elsewhere, nor is there any evidence for such a feat having ever existed' (J. H. Harvey, 1971, 104). With this roof Master Herland designed, and built, far larger than any available oak trees permitted, and he exploited the possibilities of 'built' components to an extent, and with a degree of mastery, that had been and still is unrivalled.

A change in the framing of timber floors during this period is illustrated by those of the Beaufort Tower in Winchester. Here the joists were fitted by central tenons with housed soffit shoulders (Fig. 295), a form of the joint used for joist ends that was to continue in use until the middle years of the Tudor period. The timber floors of Chichele's Tower at Lambeth Palace (Fig. 297) were a direct product of the Perpendicular style, and one that was not necessarily considered desirable by the carpenters involved. The prevailing taste in mouldings required grouped series of rolls to be applied to chamfer planes, which in the case of mullions or floor joists induced the use of knife-edged cross sections, a form that was particularly weak for a joist subjected to sagging stresses. The mechanically weaker sections led to great difficulty in designing the right-angled joints between bridging- and common joists; and, as illustrated, a combination of scribed shoulders and a tenon at the only point wide enough for that purpose had to be chosen. This was one of the earliest answers to the problem known to date, and one of the first examples which demonstrated that adherence to a fashion in visual style could, and did, conflict with mechanical efficiency.

The two large barns described are both of uncertain date since there is no definite evidence concerning their building, but they serve to illustrate the subtle way in which something more than mere conformity to visual style was achieved in functional structures to which ornament—beyond such features as verge boards, or their finials—would have been inappropriate. The timber tower at Blackmore illustrates the same point: it gives a restrained expression to the style, which must be recognised without the more obvious indications given by mouldings, capitals and bases, or doorways and windows. This point is emphasised by a comparison of the two barns, since the Widdington cross section in Plate XI gives an overall impression of the Curvilinear Decorated style, which is probably valid evidence for an earlier date ascription.

The cathedral roofs ·described do not illustrate the combination of various roof designs as clearly as do the selected parish churches, but the curious complexity of the Winchester presbytery example evidently points to similar works at cathedral level, and it must be assumed that none have survived. The continuation of two traditions, however, is illustrated; the compassed type by the roof of the choir of Carlisle, and the low-pitched type by that over the monks' dormitory at Durham. Innovations also occurred, as is shown by the remarkable nave triforium roof formerly existing at Norwich Cathedral, and the triforium roof illustrated from York Minster. The cloister roof at Hereford provides an example of a king post quite unlike the earlier ones at St. Augustine's Abbey at Canterbury, whilst the king posts of the Council Chamber at the Tower were suspended, and presaged those to be used at the Middle Temple and later, by Wren, at St. Paul's.

The three door leaves described also illustrate some slight conflict between visual style and constructional conveniences; this is apparent in the Charterhouse example, where the humped-plank surfaces had to be chamfered flat if the pierced tracery was not to be carved to a similar contour, which would not only have been difficult and expensive, but irreconcilable with the straight edges of the framing. A sound and visually pleasing compromise between style and construction was achieved by the craftsmen responsible for the door leaf of Tolleshunt D'Arcy church in Essex, where flat planks were contrasted with the moulded v-sectioned muntins and edge framing. The highly contoured plank surfaces from Fobbing church were designed to produce an interesting surface that did not require the addition of any further items, such as muntins; and it was from this type that the more restrained 'creased' doors of the Elizabethan period were evolved. The Fobbing type itself probably derived from the tradition of 'linenfold' panels, by the ends of the folds being left uncut. The iron hinges on the Fobbing door are of an earlier date and were designed for a flat surface, and the vertically moulded type of door gave rise to a change in hingeing, the ironwork being redesigned so as to fasten to the rear face of the harr durn (as shown in the drawing of the Tolleshunt example).

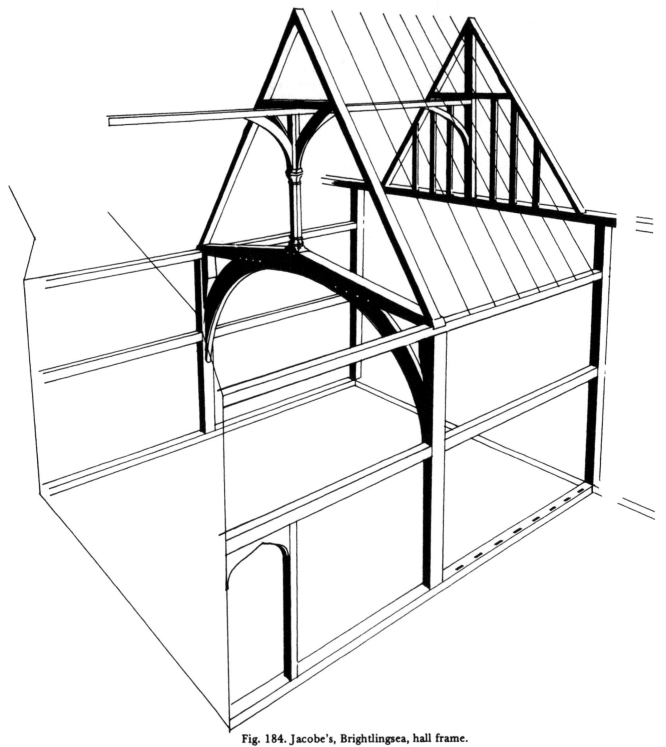

Fig. 184. Jacobe's, Brightlingsea, hall frame.

206

Chapter Seven

Examples from the Tudor Period (*c.* 1450 to *c.* 1550)

Jacobe's Hall, Brightlingsea, Essex

A late, and urban, example of the H-plan house built with an open hall which was, within a few decades, laterally divided; it may have been an inn at the time of building, because of its position in an outlying Cinque Port with maritime trade. It is of one construction and the frame of the hall is shown in Fig. 184, and the profiles of its crown post in Fig. 354: these, according to Forrester, suggest a date between *c.* 1460 and 1470. The framing of the service wing is shown as Fig. 185, illustrating the increased lengths such cross wings had attained towards the end of the popularity of this type of house.

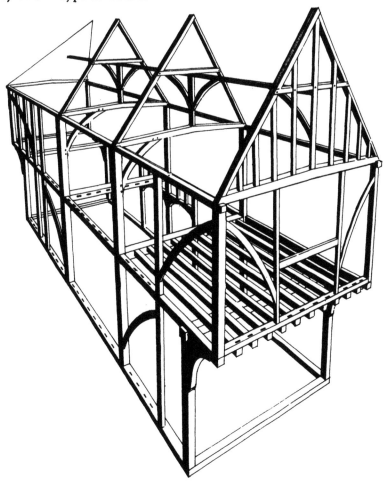

Fig. 185. Jacobe's, frame of service wing.

Table Hall, The Precinct, Peterborough Cathedral

This was the hall in which the vicars choral ate together, as is recorded in the episcopal visitation of Bishop Scambler in 1567, when the canons were actually taking dinner in the hall. This building has a ground storey of masonry, upon which is mounted a two-bay timber hall that is long-wall jettied. The principal-couple has two collars, into the lower of which are tenoned two arched braces securing spur ties—the 'legender-Stühl', or lying frame, of German carpentry. The first floor is framed with central tenons with housed soffit shoulders, and the bridging joist moulded with double ogees; the head of the ground-storey wall is cut to a great-casement profile, with fleurons. These features together suggest an early- to mid-15th-century date for the building, which is illustrated in Fig. 186.

Fig. 186. Peterborough Cathedral, roof of Table Hall.

208

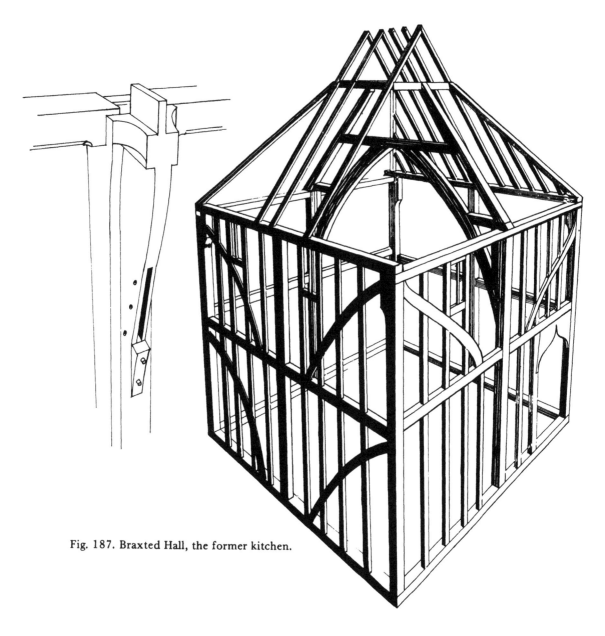

Fig. 187. Braxted Hall, the former kitchen.

The Medieval Timber Kitchen, The Hall, Little Braxted, Essex

This building stands within the moated enclosure of the former hall at Little Braxted, the present building occupying a site at some distance from it. The framing is illustrated in Fig. 187. It measures approximately twenty-two feet square and is about fifteen feet high to the eaves, and was fitted with an elaborate doorway and numerous windows beneath the eaves line. The roof is of the arched-to-collar type, with spur ties to the top plates, and dates to the closing decades of the 15th century. The interior is heavily encrusted with soot, apparently from the cooking fires, the site of which has not been established at the time of writing. The kitchen allegedly converted for use as a dovecote, presumably when the house was built further away and internal cooking commenced; but no structural evidence supports this claim. The storey posts are chamfered in two orders, as illustrated.

Morton's Tower, Lambeth Palace, London

It is stated that Cardinal Morton initiated the building of the red-brick gateway 'between the two older towers which he restored' in *c*. 1490 (D. Mills, 1956, 8). The bricks are laid in English bond insofar as the lengths between returns admit of regularity, and diaper patterns of blue-glazed headers survive on the south-west front. Avoiding speculation as to the extent and precise nature of Morton's building operation, it is acceptable that the oaken second floor of the rooms over the archway is attributed to him, and datable to *c*. 1490. The underside of this floor is, of course, the ceiling of the room below, which was one of the apartments in which the cardinal lived. All the joists were worked to highly decorative cross sections having knife-edged soffits with numerous roll mouldings (Fig. 302). The bridging joists were butt-cogged and scribed, while the common joists were fitted only by elaborate scribings. At the time of writing this is the only timber floor known that relies solely on scribed abutments.

The Gatehouse Doors, St. Martin's Palace, Norwich, Norfolk

The larger of the two pairs of entrance doors here is illustrated in Fig. 188. This example is framed in heavy plank, two planks thick and crossed in direction, the whole structure being fitted into a strongly-built surround and roved and clenched with iron. These doors were dated to between 1446 and 1472 by Repton (J. A. Repton, 1965, 34). The timber used was inadequately seasoned, and the edge joists have suffered as a result of contraction, but apart from this minor fault these doors are considered to be excellent representatives of the Perpendicular carpenters' craft. As Dr. Harvey stated (J. H. Harvey, 1961, 102), 'English architecture was tending to degenerate into lavish frivolities'—a fact which these doors illustrate well, being more a work of wood carving than of door construction.

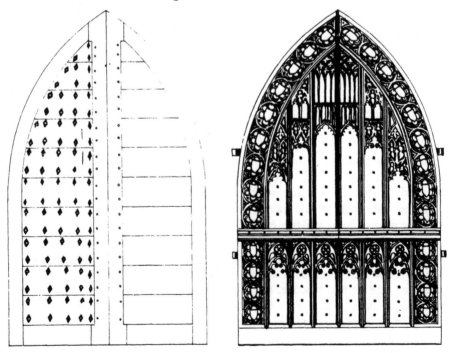

Fig. 188. Norwich, St. Martin's Palace, door leaf.

Paycocke's House, Coggeshall, Essex

This deservedly well-known building forms the frontal range of an earlier complex of various dates, and its framing is illustrated in Fig. 189. John Paycocke, it has been suggested (G. F. Beaumont, 1840, 169), built the house, before 1505, when he died, for his son Thomas and his daughter-in-law, whose initials T.P. and M.P. frequently occur on the carved ceiling timbers (floor joists). Several important developments are illustrated here, perhaps the departure from the three-part plan of medieval houses being among the most important, although in this example the difference is not marked. The front elevation is possibly the best example of a Perpendicular house front, timber-framed and jettied with oriel windows. The carving is of stylistic interest since it is definitely Gothic, with, for example, a vine trail along the fascia of the jetty, and cusping of a crisp quality on the joists' undersides. The material prosperity of this family postulates the employment of a leading medieval architect and carpenter, as is evident from the style of the design, and such a man must have been well aware of the most efficient timber joints of his times; even so, just a few years before King's College chapel roof was finished, he used the tenon with housed soffit shoulder (Fig. 295).

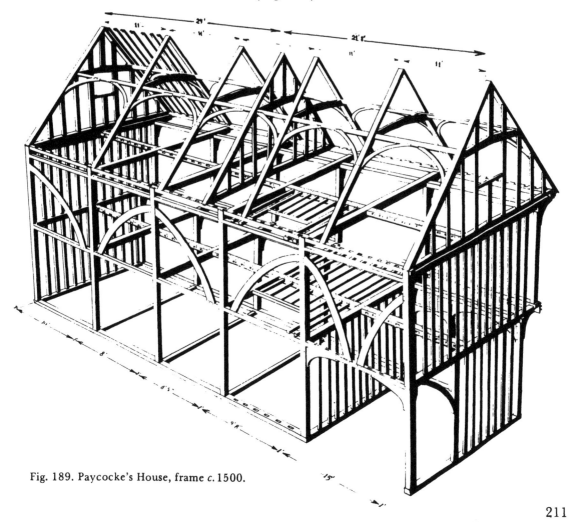

Fig. 189. Paycocke's House, frame *c.*1500.

The Old Hall, Lincoln's Inn, Chancery Lane, London

According to Sir J. W. Simpson, K.B.E., the Society of Lincoln's Inn was paying rent to the Bishop of Chichester in 1422, and had been previously situated on the south side of Holborn at Staple and Barnard's Inns (Sir J. W. Simpson, 1928, 19), where stood also the 'Lyncolnesynne' of Thomas de Lincoln, King's Serjeant, who occupied it from 1331. The Society, therefore, moved to the present site in or a little before 1422, where they occupied the great hall of the Bishop of Chichester which served the Society's purposes until 1489-90, when it had fallen into such disrepair that an order for its demolition was made. Sir William Dugdale's *Origines Juridiciales* of 1666 states that rebuilding was not begun until 1493, went slowly, and was finished in 1507; but the records of the Society indicate that the old building was still in use in 1494, and the new one finished in 1492, the year Columbus discovered America. This hall is 60ft. long and 32ft wide. In 1624 it was extended by a further bay at either end. The linenfold panelling is said by Simpson to be of the same design and possibly the same hand as that in Cardinal Wolsey's contemporary antechamber at Hampton Court. The great screen, dated 1624, was made by Robert Lynton, joiner, and cost forty pounds. The three great transverse roof frames are older than the existing roof, and at least one of their supporting posts is embedded in the existing brick walls; they are scissor-braced, and are undoubtedly remnants of the Bishop's hall previously upon this site. One unit of the present eaves framing is shown in Fig. 190, and the mouldings of both the early scissor couples and the later timbers are illustrated and described in the section on mouldings.

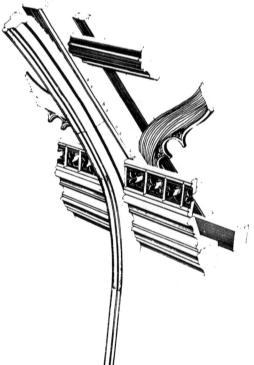

Fig. 190. Old Hall, Lincoln's Inn, framing of eaves
angles.

212

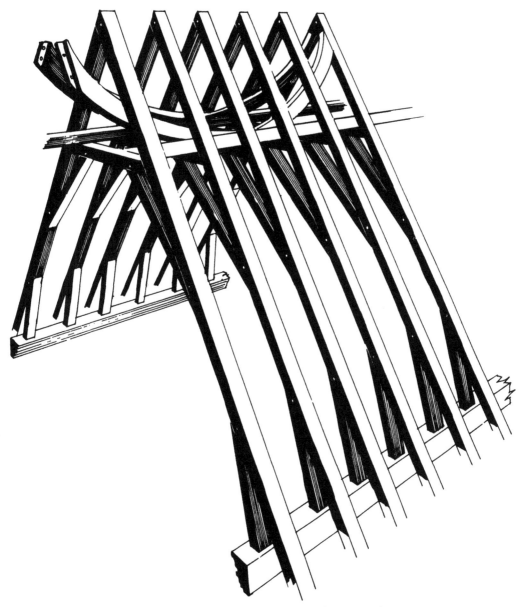

Fig. 191. Layer Marney church, chancel roof.

The Chancel Roof, Church of St. Mary the Virgin, Layer Marney, Essex

This church is suspected to have been rebuilt by the first Lord Marney between 1505 and 1510 (Sir N. Pevsner, 1956, 237), and the chancel roof is a good, if late, example of hybrid roof types. This roof has both side purlins and common collars with inclined ashlar pieces, plus the extra novelty of curved wind bracing above the purlins (Fig. 191), which probably derived from similar bracing at Wells.

213

The Bishop's Palace, Fulham, London

The information sheet, which contained neither title nor author's name, and was available during this building's vacant and redundant years, stated that it had been in the possession of the Bishops of London since the 7th century, and had also been the site of a manor-house since then. The existing buildings, built on a court-yard plan with a gatehouse, can be attributed to Bishop Fitzjames between 1506 and 1522. The bell tower was built during the time of Bishop Juxon, and is dated 1636. The Great Hall was begun in the reign of Henry VIII by Bishop Fitzjames and finished by Bishop Fletcher, between 1595 and 1597.

It is probable that the roof of the hall dates from Fitzjames's period, since such a large building is unlikely to have been left without a roof during the possibly protracted years of its completion and furnishing. This roof was at one time open and was built between two parapets and its gables, which were of red brick. It was fitted with five heavy tie beams, side purlins and crown posts without any lengthwise connections. Half the length of the roof comprises Fig. 192, in which the curious elbowed timbers of the two intermediate couples and the equally well-chamfered crown posts are of interest. At the date of original completion the roof was partitioned at its central couple for reasons not known, all of its timbers being clean and free from soot encrustation. The elbowed canting struts were secured by free tenons (as illustrated to the right of the drawing), and the side purlins were fitted by tenons with soffit spurs instead of diminished haunches, implying a date for this design prior to the introduction of the latter in *c*. 1510. This is clearly possible since Fitzjames's episcopacy commenced in 1506. The use, in the central transverse frame, of paired queen posts which clasp the scissored braces and raking struts may relate to the same practices noted in the south transept roof at Canterbury Cathedral, the roofs over Sherborne Abbey, and the roof above the north transept of Winchester Cathedral.

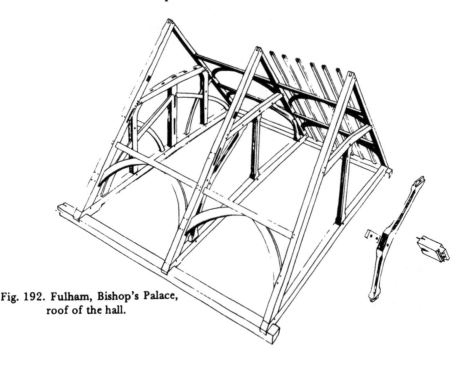

Fig. 192. Fulham, Bishop's Palace, roof of the hall.

214

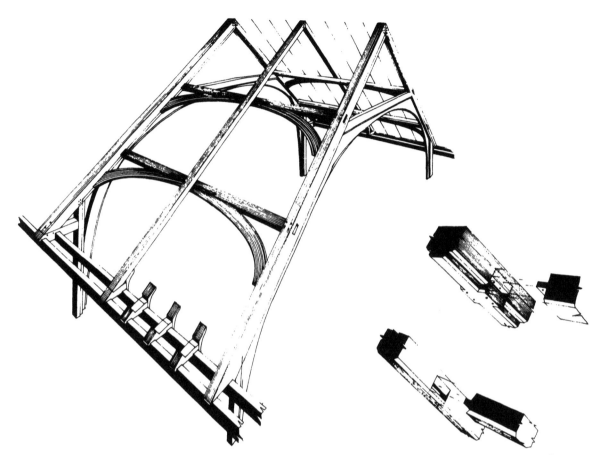

Fig. 193. King's College Chapel, the roof.

Fig. 194. King's College Chapel, side- purlin joints.

The High Roof of King's College Chapel, Cambridge

The bay length of roof illustrated in Fig. 193 is representative of the roofs above the vaulting of King's College chapel, which is a supremely important work containing the earliest examples of tenons with diminished haunches, and they are closely dated. As shown, the roof is arched to its collars with two side purlins in each pitch, and wall posts standing in the vault pockets. This is not a good roof in principle, because there was no real need for it to clear the vault's crowns, and it has required subsequent tying with steel. As at Wells, there was a break in the building here which affords a date for the innovation mentioned. The five bays of roof over the east end were built by Martin Prentice, and it is known that he drew the design in or a little before April 1480. The purlins in this part were jointed by unrefined tenons, and the two chamfered orders of the principal-rafters returned by masons' mitring. After long delays the western range of the chapel was added, under the direction of Master Richard Russell, who since 1490 had been in charge of the carpentry at Westminster Abbey. He preserved the visual appearance of the roof exactly, but fitted the purlins with tenons and diminished haunches; the main frames were completed in the middle of August 1512 (J. Saltmarsh, letter to author, 1973). The two different purlin joints are shown in Fig. 194.

215

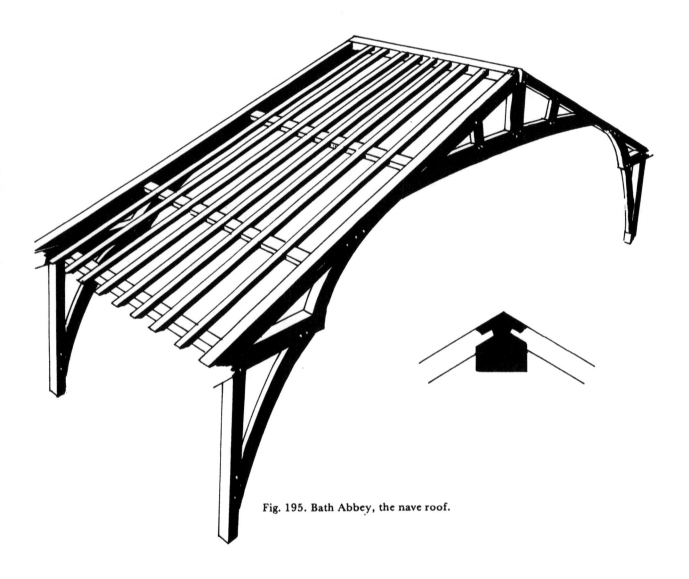

Fig. 195. Bath Abbey, the nave roof.

The Nave Roof, Bath Abbey

One bay's length of this roof is illustrated in Fig. 195, providing another example of the persisting tradition of low pitches. The design approximates to a single hammerbeam roof, and one must assume that it was designed by the brothers Robert and William Virtue, who were responsible for the entire fabric. The known dates for the nave are 1501–39 (J. H. Harvey, 1961, 117). The jointing of the rafters into the ridge piece was effected by means of tenons with diminished haunching, as is illustrated at lower right in the Figure. A probable date for the carpentry would be during the 1520s, long before the vaulting was undertaken. Of 'the vaulte devised for the chancelle there shall be noone so goodely neither in England nor in France' declared the Virtues, but their timber high-roof does not match their fan vaults in ingenuity, and it may not be correct to ascribe both carpentry and masonry to the same craftsmen.

216

This building forms an L-plan and includes the Council Chamber previously described; but the range aligned east to west that has four facade gables is now discussed. This range was completed in 1528, and was built for the resident Lieutenant of the Tower.

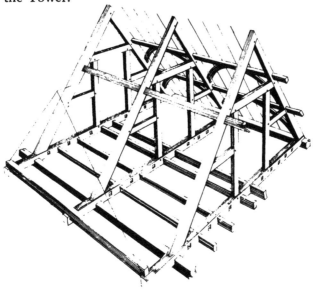

Fig. 196. The Queen's House, Tower of London, roof and attic-floor joists.

The roof is of much interest and is queen-posted, with two side purlins in its rear slope and one in the front slope, to which the facade gables are attached; the attic floor is jettied, as are those beneath it. Two bays of the roof are illustrated in Fig. 196, which shows the jettied tie beams that support the facade gables. These beams are cambered, as was traditional in open roofs, which in combination with a flat attic floor produced the anomaly shown in Fig. 197—tie beams that rise at their centres above the floor boards. The common joists are undoubtedly the originals and are, at the time of writing, the earliest examples of the deep section, measuring 9ins. deep by 3ins. wide, and fitted, of course, with diminished-haunch tenons. The windows in the gables are also important, because at this stage (1528) they have depressed four-centre heads in the 'Tudor' style and their moulded components are of the ogee-, return- and hollow profile. This last fact narrows the time lapse betwixt Tudor styling and the impending taste for square-window voids and ovolo mouldings, which had established themselves by the 1850s in the capital. The verge boards of this house are mainly the originals, and a representative apex and valley conjunction are shown in Fig. 382.

Fig. 197. The Queen's House, Tower of London, humped tie beam, flat floor.

St. Aylott's, Saffron Walden, Essex

This large house on a moated site is evidently associated with the Benedictine Abbey of Walden, which was founded as a priory in 1136 by Geoffrey de Mandeville, and converted to an abbey in 1190. 'William More, the last Abbott, did, on the 22nd March 1537, surrender this abbey. And King Henry VIII granted it, the 14th of May 1538, to Sir Thomas Audley' (P. Morant, 1816, 548). Little else can be learned of its origins but it is of too high a quality to be considered in any but the monastic context, and its date must, in view of its jointing and decorative details, lie between *c.* 1512 and the surrender of the Abbey in 1537. The house is of a rectangular plan and has a red-brick ground storey, laid in English bond, and a jettied timber-framed first storey incorporating many mouldings that confirm the proposed dating; it is roofed with the elaborate system of framing illustrated in Fig. 198. This roof was initially fitted with an attic floor, which survives *in situ* with its boards rebated into the tie beams (as illustrated).

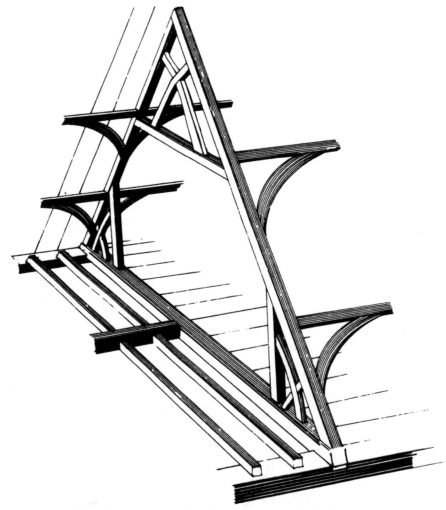

Fig. 198. St. Aylott's Farmhouse, a principal-couple.

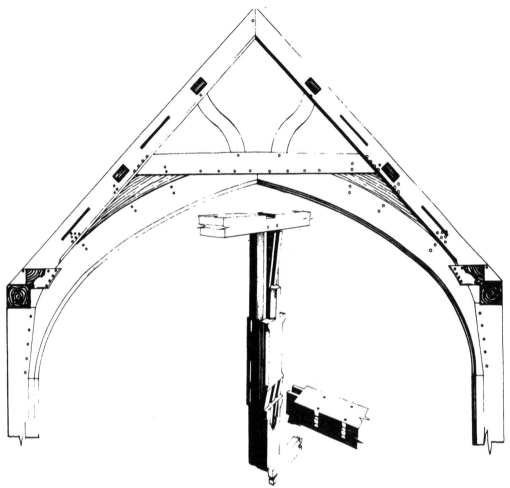

Fig. 199. Chingford, the Greate Standinge, section through roof.

Queen Elizabeth's Hunting Lodge, Chingford, Essex

This building was formerly designated the 'Greate Standinge', and was designed and built by 1543 as a necessary adjunct to forest or parkland hunting. A warrant exists that was issued by King Henry VIII, authorizing the payment of £30 to George Maxey, carpenter, for the 'fynyshinge as wall of(f) on greate stondeinge', probably indicating the completion date for the structure.

It is of a rectangular plan, and is aligned roughly east to west. It is wholly framed in timber, and comprises three bays and three storeys with an attached stairs tower, producing an L-plan. The first floor was fitted with unglazed windows for viewing the hunt, and is cambered, in the same way as a ship's deck, to run off such rainwater as must have entered during wet weather: the scuppers, however, are not visible today. The roof was always open to its apex and is of the arched-to-collar variety. The arch timbers are affixed to the rafters by several free tenons (as shown in Fig. 199), and the top plates are secured by short spur ties. The spandrels of the arches were filled by fitted planks, and the rafters' apéxes are jointed by counter-bridling with numerous pegs, as at King's College. The moulding section is illustrated in Fig. 372, and the common joists of the upper floors at the centre of Fig. 199; these are, at the time of writing, the earliest known examples of their kind—pairs of single tenons, each with diminished haunches.

219

The Great Hall, Middle Temple, London

'This Hall was begun to be built about the year 1564, being the third year of the Treasurership of Edward Plowden esq., who had the whole direction and management of that work, . . . as appears by an Order of Parliament 7 May 1567. When Thomas Andrews Esq., was chosen Treasurer, reserving to Mr. Plowden the continuance of the care and direction of the building of the New Hall, which he finished, and all of its ornaments, about the year 1571' (Lingpen, 1910, 104).

The roof took at least ten years to build and Plowden was 'procurator promoter for building the new Hall and making collections' towards its costs. These funds took the form of compulsory levies from members of the Inn. The date of absolute completion appears to have been 1573, and the official opening had previously taken place in 1570. In 1562 Plowden wrote to Sir John Thynne requesting the loan of the head carpenter from Longleat, where a single hammerbeam roof also exists. The hall's dimensions are: length, 100ft., width, 40ft., height of walls, 29ft., and height to ridge 59ft. This carpentry is extremely important in view of both its location in the capital and its firm dating, and the moulding profiles worked on the roof timbers are shown in Fig. 375, in which one *turned* pendant is also shown.

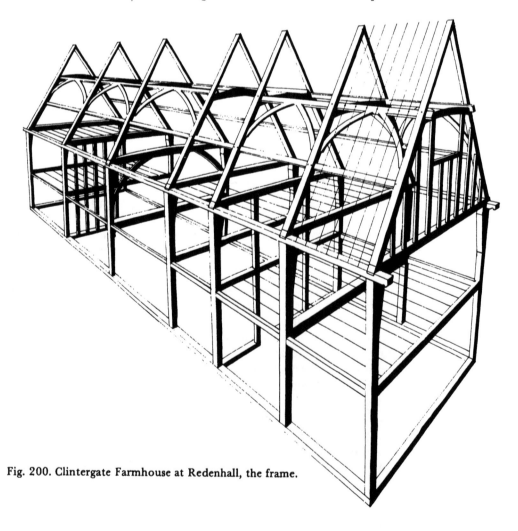

Fig. 200. Clintergate Farmhouse at Redenhall, the frame.

220

Clintergate Farmhouse, Redenhall, Norfolk

A timber-framed house of two storeys and five bays, between the second and third of which is a framed chimney bay containing a red-brick chimney-stack, which dates from the first building operation. All of the storey posts are jowled, the walls are side-girt, and the roof has side purlins with curved in-pitch braces beneath them: the collars are well cambered and act in compression.

The five bays of this plan are reminiscent of Paycocke's House in Cöggeshall, Essex, which dates from between *c.* 1495 and *c.* 1505. Two of these bays (those adjacent to the stack and west of it) provided a definite parlour-style room which was fitted with carved and decorative storey posts. The same two bays above this and on the first floor had been designed as an open chamber to the ridge of the roof. The stairs originate from the first building operation and were contrived as winders at the rear of the chimney-stack. The top plates are scarfed in the ultimate form, that is face-halved and counter-bladed and provided with six edge pegs. The frame of the house is shown as Fig. 200, the floor joists' end joint as Fig. 295, and the scarf used as Fig. 271. The modifications that were effected during the long working life of this house have been more fully published before (C. A. Hewett, 1972, 61–64).

Doe's Farmhouse, Toothill, Essex

A good specimen of the two-bay house with intervening chimney bay, which retains both its original stack and also, on the ground floor, its two different types of fireplace, one arched for the parlour, the other beamed and very wide for cooking purposes. The entire frame of this buiding is shown in Fig. 201.

The whole structure was encased in red brick, perhaps during the 18th century, an undertaking which served to hide the jetties contrived at both the terminal tie beams. A handsome architectural feature of the house is its stairs tower, which was integrally built against the rear wall and contains its original newel stair and ovolo mullioned windows. This is shown both in the general drawing, and as a separate entity in Fig. 202, which indicates the fitting of both treads and riser planks to the outer frame. The newel top is shown inset. A door leaf that survives there is also shown on the left of Fig. 203; it illustrates the arrival of the shadow-planed or 'creased' door type, which was to remain invariable until the advocation of leaves that were framed around fielded panels during the 18th century. Later examples do not have the rear framing of this early specimen, nor its elaborate ride hinges.

221

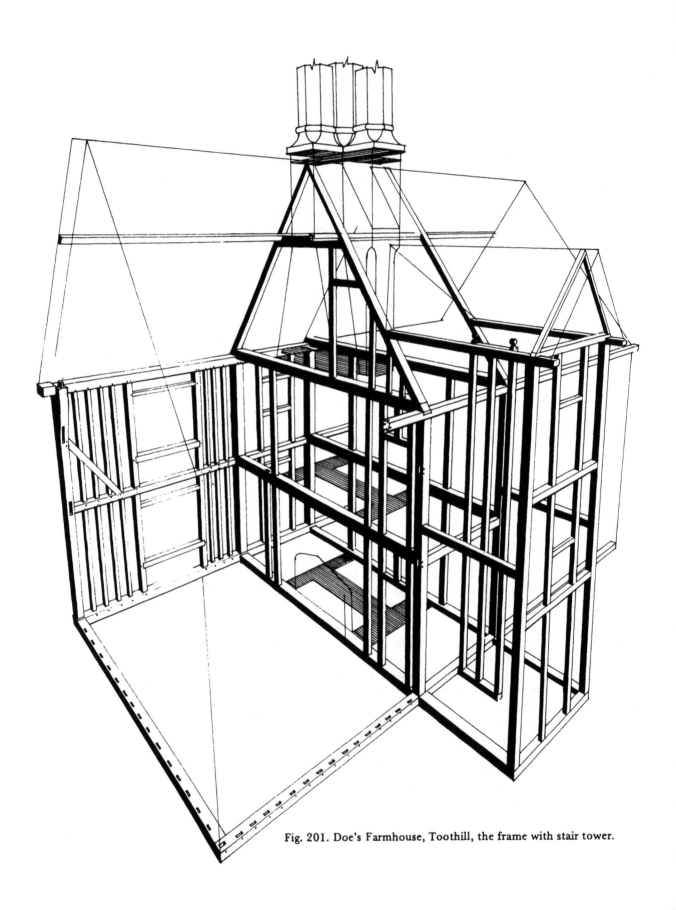

Fig. 201. Doe's Farmhouse, Toothill, the frame with stair tower.

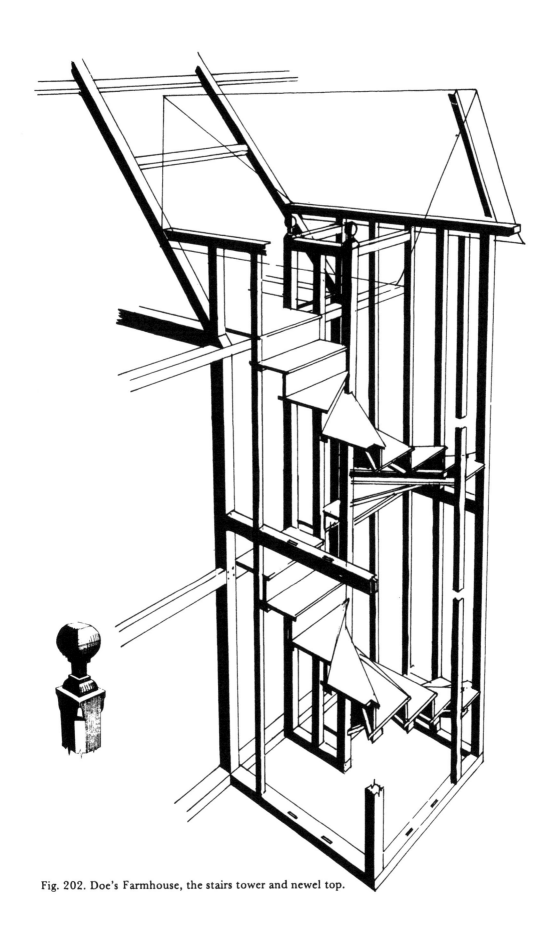

Fig. 202. Doe's Farmhouse, the stairs tower and newel top.

Door Leaves from Doe's Farmhouse, Toothill, Essex

This is a handsome specimen of the door type that was emerging and was to be unchanged during the 17th century, as a creased door, a term denoting the vertical mouldings that were designed to create a shadow. This example is early enough to retain framing, which is applied to its front, and also a form of strap-hinge ride. The plainer specimen was probably intended as an internal one, and the stronger framed door seems to have been intended for the front entrance (illustrated, with cross sections on the right, in Fig. 203).

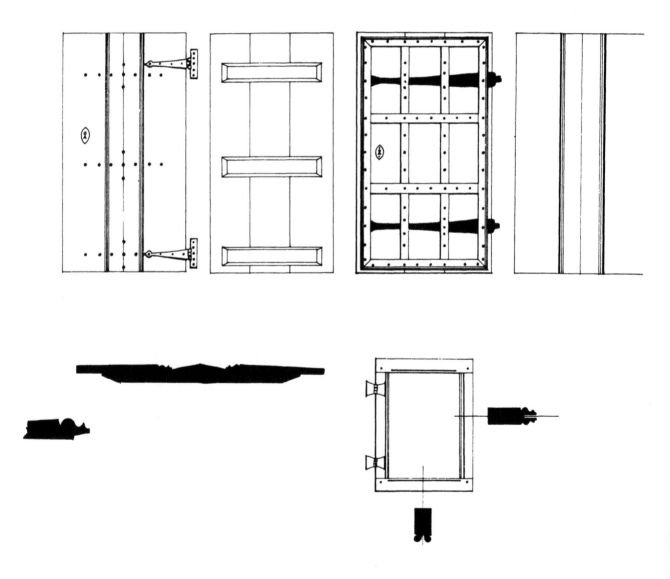

Fig. 203. Doe's Farmhouse, selection of door leaves.

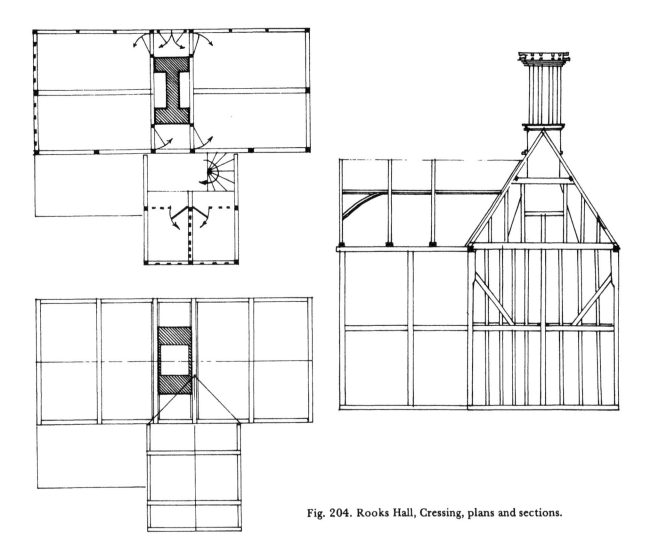

Fig. 204. Rooks Hall, Cressing, plans and sections.

Rooks Hall, Cressing, Essex

This house is dated, the visible evidence being two figures '75' carved on one of the frontal side girts; the other two can only be '15'. This appears from the front elevation to be a truly two-part design, with the central chimney bay containing a very fine tall red-brick stack that mounts four elegant octagonal shafts. It is, however, an early example of its type because no stairs were designed into its plan, and part of an earlier building at right angles to its rear wall was retained because useful stairs existed within it. It thereby conforms to a T-plan (Fig. 204).

The whole timber frame of this building is shown in Fig. 205 as a perspective; the wind bracing was reduced to the four returns above the side girts. The attic floor originates from the first building operation and is fitted between cambered tie beams. The scarf used to unite the top plates is an early example of its type (insofar as is known at the time of writing), and is illustrated in Fig. 270; it is the face-halved and bladed joint with edge pegs which was to dominate scarfing throughout the 17th and 18th centuries.

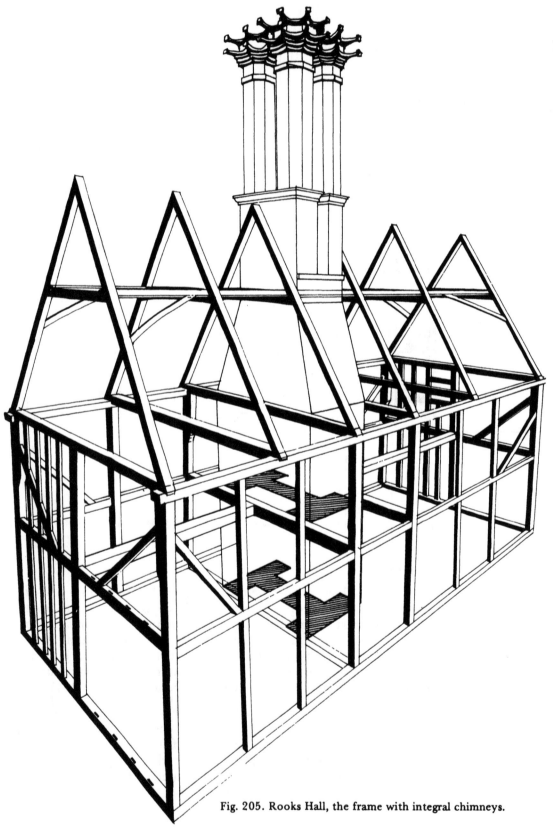

Fig. 205. Rooks Hall, the frame with integral chimneys.

The Tudor Period—Summary

The selection of carpenters' works representing this period has been chosen because the examples illustrate several major changes that affected, for example, the internal economy of houses, just as much as their external appearance. Jointing technique had, in one category, advanced to its highest mechanical efficiency, having reached its lowest level two decades previously, when structural soundness had been sacrificed to superficial appearance. The best illustration of this deterioration is to be found in the timber floors of Morton's Tower, at Lambeth Palace (Fig. 301). This was the logical development of the type used in Chichele's Tower, but no integration of the timbers was attempted, and the scribing of the tenons' shoulders was carried to its ultimate limit, producing simply scribed abutments (as shown). This work dates from the final decade of the 15th century, and although similar floors were designed in country houses, such as Jacobe's Hall in Essex, it seems that they were never so fully committed to their scribings; the mouldings of their joists were often run into a suitably scribed housing, thus combining the new fashion without chamfer stops with the earlier housed soffit shoulders. An example is illustrated in Fig. 302.

The ultimate development affecting framed floors was made during the ensuing decade by Richard Russell, insofar as the available evidence can indicate, and although it first appeared in the context of roof framing (as did the spurred shoulders of Bubwith's library roof at Wells), it rapidly affected the design of timber floors. This adjunct of the tenon was the diminished haunch, which applied the principle of the spur to the compression side of the tenon, providing for it a buttress (Fig. 303). The relative mechanical efficiency of these two forms is affected by the relationship of the joints to the neutral axes of the major timbers, which is a matter discussed on p. 286. The latter of the two was subjected to industrial testing, and the report on its behaviour under shearing stress is quoted on page 284. The roof mentioned in connection with Bath Abbey used the newer principle, and provides examples of roof designs which are nowise equal to those of earlier periods, but which incorporated better joints: they also illustrate the use of this form as a dating factor.

Two rare survivals from the period are the 'Great Standinge' at Chingford, and the timber-framed external kitchen at Little Braxted Hall, both in Essex. To say that they are rare is misleading, since it might be inferred that they were equally rare in their own time; but, on the contrary, both types must have been common. External kitchens of timber (which must have been at great risk from fire, particularly during spit-roasting) were popular, as can be assessed from the frequent illustrations of them on maps, such as the famous 'Walker ' maps of Essex. Standings for the viewing of the hunt must have been almost equally common when the wealthy manors maintained great deer parks, and an earlier, larger example than the royal one at Chingford survives at Galleywood, which was apparently designed to afford some summer accommodation as well as viewing space on its first floor. It is a feature of both standings that the first floors are cambered like a ship's deck in order to drain off rainwater, which must have entered freely through the unglazed and numerous apertures. The second floor at the Chingford standing is the earliest known to date which has joists with pairs of single tenons, each with a diminished haunch, the probable completion date being 1543. This was an important development, closely involved with the use of deep-sectioned joists, employed for greater rigidity.

The roof of Table Hall at Peterborough Cathedral is an example of what is known in Germany as the 'legender Stühl', or lying frame. This system undoubtedly derives from more direct forms of cruck-bladed roofs; its use was not infrequent in Essex, where it may be found in the context of early 17th-century agricultural buildings when a clear attic or first-floor space was needed, in granaries or malthouses.

During the early part of this period the secular large house (like the Brightlingsea example, Jacobe's Hall) continued the then long tradition of the central open hall with two storeyed cross wings. However, sweeping changes were imminent, and at Jacobe's the open hall was laterally divided soon after completion by a first floor that derived from the scribed floors in London, as at Lambeth Palace. This house must have been one of the last built to the old design, and both Clintergate Farmhouse and Paycocke's House illustrate the addition of extra accommodation and the adoption of plain rectangular ground plans in the closing decade of the 16th century. It is significant that the floor joists of these two both used the tenon with housed soffit shoulder, rather than the soffit spur, the carpenter of Paycocke's preferring stopped moulding and carving to the meretricious types of scribed abutments, or even the superior scribed soffit housings which are occasionally to be found. St. Aylott's at Saffron Walden represents the new, long but rectangularly planned type, and its floors were fitted with diminished haunches.

Doe's Farmhouse and Rooks Hall, both in Essex, represent the type which was to be colonially advocated in the New England States, comprising two bays with an intervening chimney bay and integral chimney-stack; the latter was incorporated as an architectural feature for the first time in secular domestic buildings. In the final form of these houses the stairs (of newel or winding type) were placed behind the chimney-stacks, and the front door in front of the stacks; but the selected specimens, being both early and provincial examples of their kind, represent two alternative stair systems. At Rooks Hall the service bay (part of an earlier house) was retained because it contained a stair and formed a T-plan, while at Doe's Farmhouse a handsome stairs tower was built against the rear wall, thus continuing as an architectural tradition what had been incepted as a necessity when open halls were floored. Jacobe's Hall provides a good example of such a stairs tower, which was built in red brick at the time when its first floor was intruded into the hall.

At Rooks Hall a new form of scarf joint was used for the top plates; this joint is face-halved and bladed, and could be cut mainly with saws and at low cost (Fig. 270). In the same building the final form of wind bracing appears; it is straight and primary since it was applied before the studs (which were cut and spiked to the braces), and is also confined to the angles of main frame above the side girts in each of the four corners of the house. These were the first houses to have integral attic floors, and some, like the Queen's House in the Tower of London, retained through the strength of tradition cambered tie beams which, combined with flat floors, produced obstructions for the pedestrian at bay intervals! The Queen's House can be dated, having been built for Henry VIII and finished by 1528. This building had another novel architectural feature—the facade gable, which, in pairs on front elevations, became invariable throughout England and New England in the following century. The first known use of floor joists fully three times as deep as they were wide, fitted with diminished-haunch tenons, was in the framing of these floors.

The doors in St. Martin's Palace at Norwich illustrate the extent to which carving was lavished on such items in this period, their sole fault being the inappropriate use for this purpose of green timber in two crossed layers; if economy was the reason for this it is surprising, since the building was associated with the Cathedral.

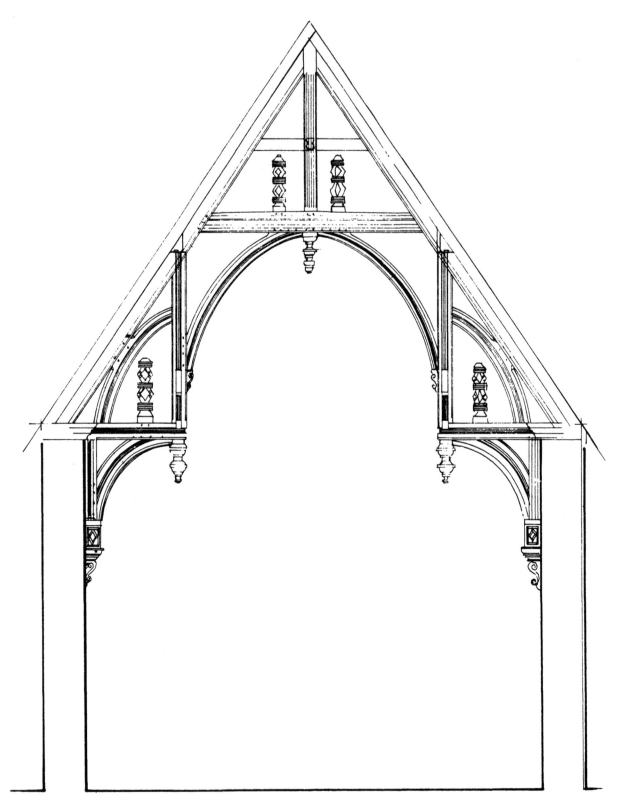

Fig. 206. Staple Inn, section through great hall.

Chapter Eight

Examples from the Renaissance and after (1581 to 1890)

Staple Inn, High Holborn, London

According to the information published by the present occupants of these buildings 'Staple Inn has a long history of occupation by lawyers, but its earlier history of use by merchants is hidden' (M. E. Ogborn, 1964, 6). Nearly three hundred and fifty years ago, according to the cited source, Sir George Buck wrote: 'Staple Inn was the Inne or Hostell of the Merchants of the Staple (as the tradition is) wherewith untill I can learn better matter, concerning the antiquity and foundation thereof, I must rest satisfied'. More is known today concerning the history of the site and the buildings formerly upon it, the first of which was designated 'le Stapled Halle' in a document of 1292, but the existing hall and residential range are of relatively recent dates. The great hall was built through the subscriptions of the Fellows, in 1580-1, when Richard Champion was Principal; his heraldic achievement was placed on a carved timber corbel above the oriel window. The Principal in 1585 built the houses on the west side, and, in 1586, the row fronting on to Holborn. The authenticity of these buildings has frequently been questioned, in view of which a summary of their recent vicissitudes is justified.

At about 7.30 a.m. on Thursday, 24 August 1944, a German flying bomb made a direct hit upon the great hall, and a photograph in the library of the present occupants shows the hammerbeam roof, virtually intact, surmounting a heap of brick rubble that had been the walls. This was taken within minutes of the event. The stained glass had been previously removed for safe storage, and was reinstated after the hall had been rebuilt 'on its original site and as nearly as possible to its former design; the stained glass windows are again in their old positions. One of the roof trusses has been reconstructed from the original oak, and the carved pendants and features on the new trusses are almost entirely the originals' (*Institute of Actuaries Year Book*, 1973-4, 14). After a close examination the writer considered that this was an understatement, and saw no reason why the moulded components of this roof should not be the originals, reassembled; if not, their production and patination represent unrivalled antique faking. The roof remains a proper object for study, and is of impressive integrity.

The latter source also states: 'In 1936, complete restoration of the old buildings in the frontage of the Inn towards Holborn became essential', and some account of what was done at that time is relevant. The architects involved were Messrs. Daniel Watney and Sons, and their files recording the progress of the work have survived. These include the first report on the condition of the buildings before work was set in hand, and this report, dated 6 January 1936, states on page three: 'we would mention that in the course of our survey we have not found any trace of comprehensive works of restoration, although here and there individual timbers have been repaired from time to time'. It is, therefore, apparent that this great range of timber-framed buildings had survived from 1586 until 1936 in virtually their original condition, and to follow the records of the restoration is essential. Another report signed by the Clerk of Works and dated 4 March 1938 records: 'sixty per-cent of old oak front picked-up and tied back to steelwork', and yet another dated 18 March in the same year reads' 'Old oak front repaired. Windows, the *repairs* to lead lights in windows well in hand' (writer's italics). In the light of these records and the fullest possible examination on site, it is considered that this front elevation is an invaluable document, giving full information concerning the methods, materials and designs of London carpenters at the close of the 16th century.

A section through the great hall is shown in Fig. 206. The roof is of a hammer-beam design jointed mainly by chase tenons and resistant only to compression; the actual hammerbeams function as tie beams, and have double lap dovetails for this purpose. The high king posts in each transverse frame do not function as pendants, and the profusion of mouldings conform to the ovolo and sunken-channel fashion of those years. The ornaments fitted as 'standards' above the collars and hammers have amused and puzzled many previous writers; no explanation is here offered, other than that of overseas trade.

The range fronting on Holborn is of more interest and comprises seven facade gables, the part now numbered 338–337 forming a separate unit which has not been internally examined. The other five-gabled range with its central freestone archway and oriel window can be described in some detail, however. The framing was of oak at both front and rear walls, as was that of all internal partition walls; the whole structure was three storeys high and jettied at each floor, with an attic room

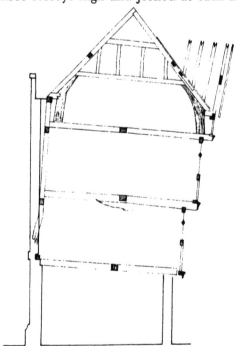

Fig. 207. Staple Inn, section through residential range, showing base crucks.

behind each facade gable, calling to mind the contemporary observation that 'Buyldynge chargydde with iotyes is parellous when it is very olde' (L. F. Salzman, 1952, 205). The truth of this statement is illustrated in Fig. 207, which is scaled from one of the architect's original cross sections, drawn before restoration was begun. The timber rear wall was at that time embedded in 18th-century brickwork, as shown, and the whole frame was leaning over the street; the jetties were out of balance, being confined to one elevation. The most important and surprising fact is the use of base crucks with spur ties to frame the partitions between each facade gable.

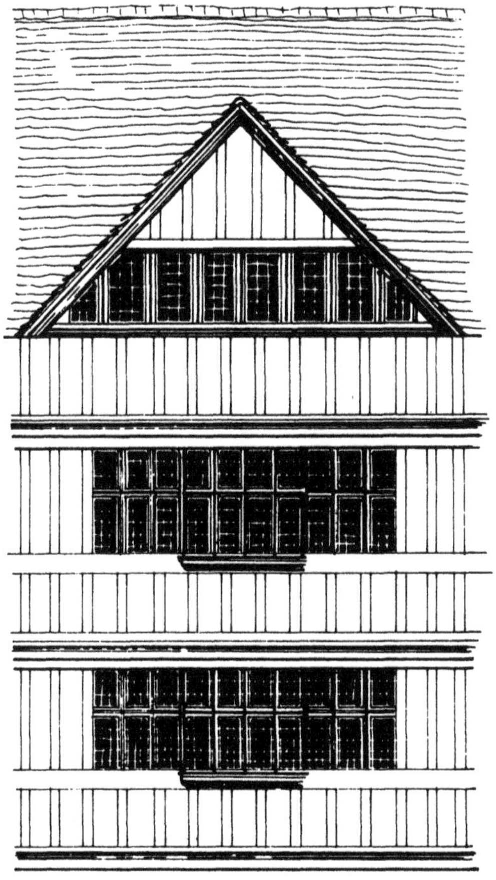

Fig. 208. Staple Inn, front elevation of residential range, the ground storey omitted.

One bay of the front elevation is shown as Fig. 208, the architectural significance of which may be considerable, since full-width attic casements are not known elsewhere in the south-eastern context at the time of writing. The bressummers used were not of the fully-framed type, but were masked by the sill timbers of the jettied floors; no wind bracing was applied to the wall frames, but the cross walls were braced to resist the imbalance of the jettied storeys. The window mullions and transoms were ovolo-sectioned, and the sets of casements were not of the 'eared' tripartite style, except for the projection forward of their central thirds of lights. Finally, the framing is not pegged at all joints, being frequently confined to every fifth stud.

House at Reepham, Norfolk

It was verbally asserted some years ago that this house was to be called Candle Court, information which may serve to identify the building. The latter comprised two bays and two storeys, with a front-wall jetty contrived in the manner known as 'hewn' jettying, which was indubitably the final form of this decorative feature. The method is known during the early 18th century in, for example, the State of Connecticut, U.S.A. The frame was wind-braced, heavily timbered, and had a queen-post roof, with the purlin braces set vertically (as shown in Fig. 209). In this case the projection of the upper floor was *cut* from the storey post, and by this means two such posts could be cut from one tree. The floor joist joint is shown in Fig. 306.

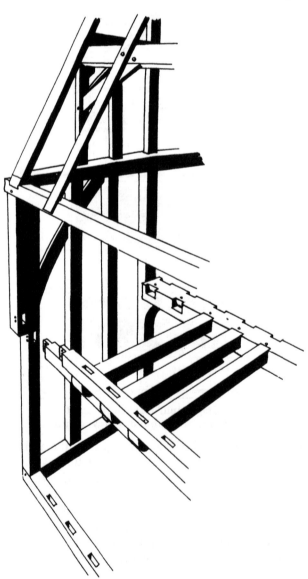

Fig. 209. Reepham house, framing, with hewn jetty.

234

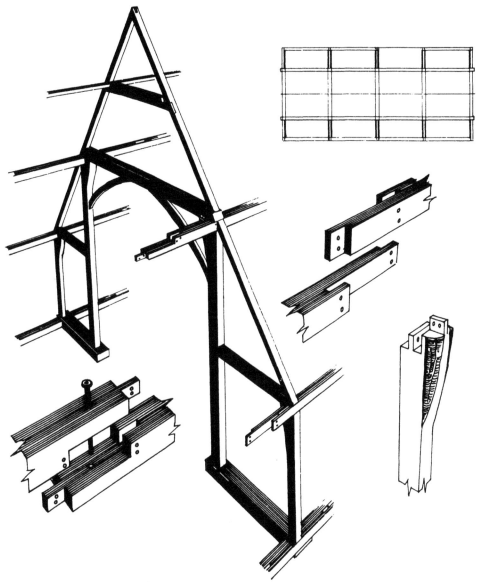

Fig. 210. Barn at Springwell Farm, frame details.

Barn at Springwell Farm, Great Chesterford, Essex

When examined this barn was partly unclad, and it may not survive at the time of writing. It was a late example with two aisles, four bays, and no midstrey, and is illustrated in Fig. 210. This gives a plan, one transverse frame, the two scarf joints used and one waney-edged jowl. The top plates made use of the edge-halved and bridle-butted scarf, and the outshut top plates the face-halved and bladed scarf, which indicates that the most trusted joint was selected for the most stressed situation, and the newer joint for the least stressed; the older scarf was also bolted with iron. A theoretical method of arriving at a probable date for this barn is to divide the time lapse between the latest certainly dated example of the edge-halved scarf (1655) and the earliest dated specimen of the bladed scarf (1575), resulting in a proposed ascription of *c.* 1615.

The High-Roof of the Nave, Sherborne Abbey, Dorset

This part of the church dates to the late 15th century (J. H. P. Gibb, 1972, 30-1), when both the choir and choir aisles were rebuilt in the Perpendicular style. The roof above the vaults (Fig. 211) cannot be the original, but appears to date from the 17th century; it is a simple queen-posted design well wrought in oak, without recourse to any form of ironwork.

Fig. 211. Sherborne Abbey, a roof frame from the nave.

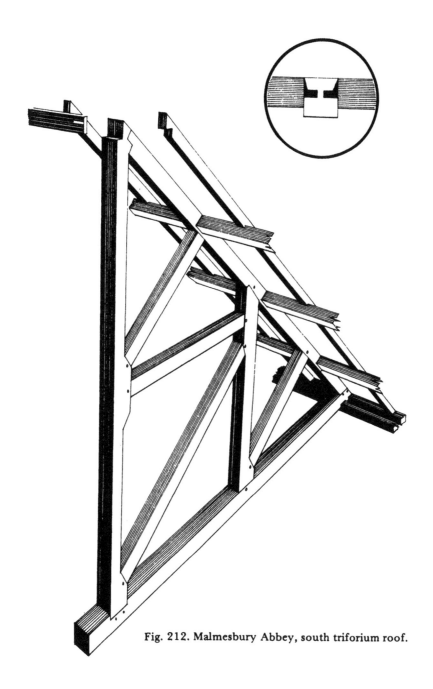

Fig. 212. Malmesbury Abbey, south triforium roof.

The Roof of the South Triforium, Malmesbury Abbey, Wiltshire

(Fig. 212.) This work was evidently designed after 1512, since the side purlins were fitted with diminished-haunch tenons (as shown in the inset diagram). A direct and impressive design, wrought in well-finished oak, it presages the roof types of the later Renaissance, and is the origin of the design used later for the southern triforium of Salisbury.

237

Deal Tree Farm, Hook End, Blackmore, Essex

This was a small farmhouse in a loosely distributed secondary settlement, representing a further intake in this forest parish of poor-quality land. If it is the first house on the site it may well indicate assarting as late as the 17th century. The framing was examined during rebuilding in 1969, when it was spoilt as an historic document, but what was then visible is shown as Fig. 213. The inset shows the short form of edge-halved scarf joint used, and the door heads that were fitted to the cross-passage openings; the type of end fenestration used must indicate a formerly open chamber, presumably at that end of the building, with a service bay beyond the cross passage.

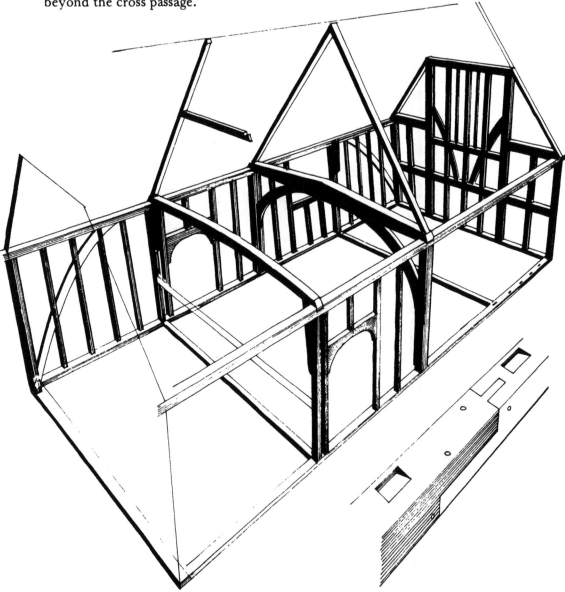

Fig. 213. Deal Tree Farmhouse, the frame.

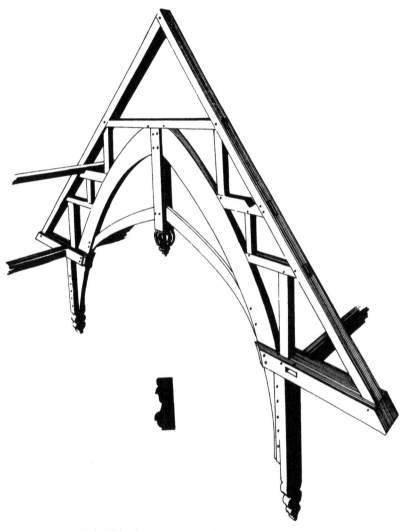

Fig. 214. Chipping Ongar, reinforcing roof couple (1643).

The Chancel Reinforcements, Church of St. Martin of Tours, Chipping Ongar, Essex

One transverse frame of this is illustrated in Fig. 214, together with the moulded section of its wall-plate fascia (shown inset). This comprises three frames, carrying side purlins designed to support the three earlier and fragmentary systems of roofing. Essentially these are collar arches but of a form that is inherently weak, being tenoned into a central pendant; conscious of this weakness, the carpenter added curved timbers in extension—the weakest method of counteracting the initial fault. Wall posts and numerous struts and spur ties complete the assembly, which is dated on one pendant 'W S — 1642', -43 or -47 (it is difficult to determine).

239

The Nave Roof, Dore Abbey, Hereford

Originally a Cistercian Abbey founded in 1147 by Robert, a son of Ewyas and grandson of Ralph Earl of Hereford, it was suppressed in 1535 (F. C. Morgan, 1967, 3). The buildings fell into ruin for many years and were subsequently restored by John, 1st Viscount Scudamore, in 1632 (F. C. Morgan, 1949-51, 163). By that time it had degenerated into a shelter for cattle. The roof was rebuilt in local oak by John Abel, who was recorded as an 'architector', the stone vaults were not replaced and a timber roof was built with an attached ceiling: 204 tons of timber were used at a cost of five shillings per ton, Abel felling the trees and Scudamore transporting them to the site. Enough reused timbers from the pre-Dissolution roof survive in the triforia to make a hypothetical reconstruction possible, but the existing 17th-century example is of much interest, although it cannot be thought to represent any continuing tradition relating to the former roof.

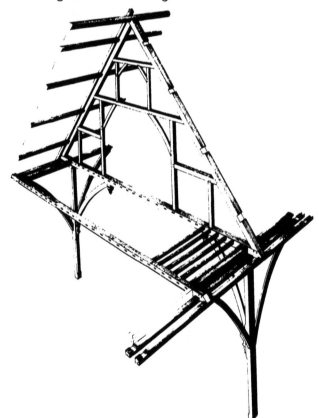

Fig. 215. Dore Abbey, the roof by John Abel.

The bays contain 10 common rafters each, double wall plates and straight tie beams spanning 32ft.; all timbers are straight and four side purlins and a ridge piece were fitted. This is illustrated in Fig. 215, in which the three tiers of vertical struts with straight bracing may be studied. The wall posts beneath the ceiling with their pendants and scrolled corbels show clearly that the Gothic decorative forms and devices were dead (Fig. 379).

240

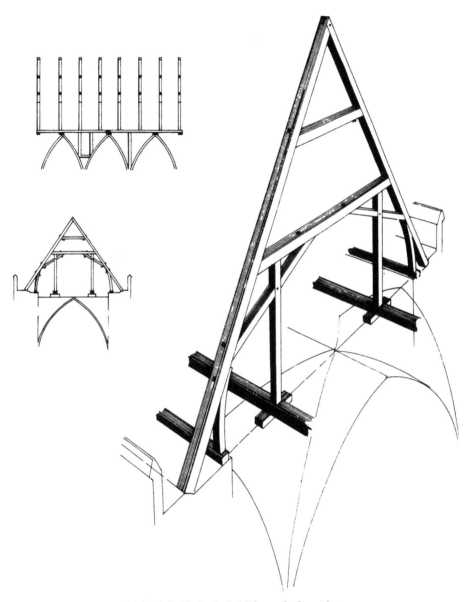

Fig. 216. Wells Cathedral, high-roof of north transept.

The High-Roof over the North Transept, Wells Cathedral, Somerset

One couple from this, the latest of the roofs at Wells, is shown in Fig. 216. This work seems to date from 1661, when a sum of £227 (in round figures) was expended upon 'timber and workmanship about the church' (L. S. Colchester, letter to author, 1972). The roof was planned to stand on the transverse vault ribs and is of a strange design with queen posts, soulaces and compassed ashlar pieces. Some timbers were evidently re-used, and are of oak, the majority being of larch. This design was clearly influenced by that which resulted from the conversion of the nave high-roof to parapets, and the bolts used have screw-threaded nuts which, if original, precede Wren's use of them at St. Paul's Cathedral by almost half a century.

241

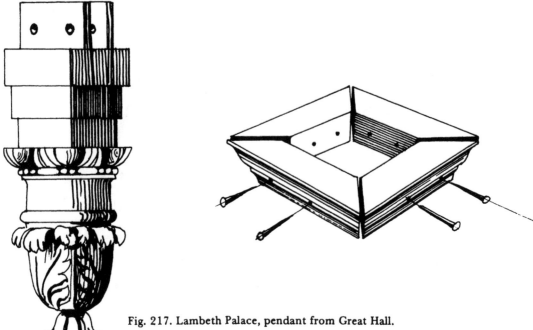

Fig. 217. Lambeth Palace, pendant from Great Hall.

The Great Hall, now the Library, Lambeth Palace, London

The earlier hall was almost destroyed during the Commonwealth. Upon the Restoration in 1660 William Juxon was appointed Archbishop and instituted its rebuilding. In his will he specified: 'If I die before the Hall of Lambeth be finished, my executors to be at the charge of finishing it according to the model made of it, if my successor shall give leave'. Despite the advice of friends Juxon was determined to build it in a Gothic style, and was so successful in this that Samuel Pepys in his *Diary* records his visit to the building in 1665, describing it as 'a new old-fashion Hall'. The open timber roof is not illustrated; it is a complex timber frame with both hammerbeams and arches to its collars, a medieval roof to which Renaissance decorations were applied. One pendant carved with acanthus leaves is shown as Fig. 217, in which one of the numerous mitred and spiked-affixed square collars is also illustrated. The predominant moulding can be seen in Fig. 217.

The High-Roof of the South Transept, Lichfield Cathedral, Staffordshire

A bay of this roof, from the northern end of the transept, is illustrated in Fig. 218, with the tie-beam seating joint shown enlarged at lower right. This is the roof type used throughout the general restoration under Sir William Wilson from 1661-9 (J. H. Harvey, 1961, 139). It shows some variations, mainly of jointing technique. Most of the eaves triangles are bolted with forelock bolts, and the lean-to roofs are of re-used timbers that might make reconstructions of the originals possible. Many of these roofs were fitted with common collars in addition to side purlins, all were made from very heavy oak (much heavier in fact than was necessary), and none has any merit that compares with earlier cathedral roofing.

242

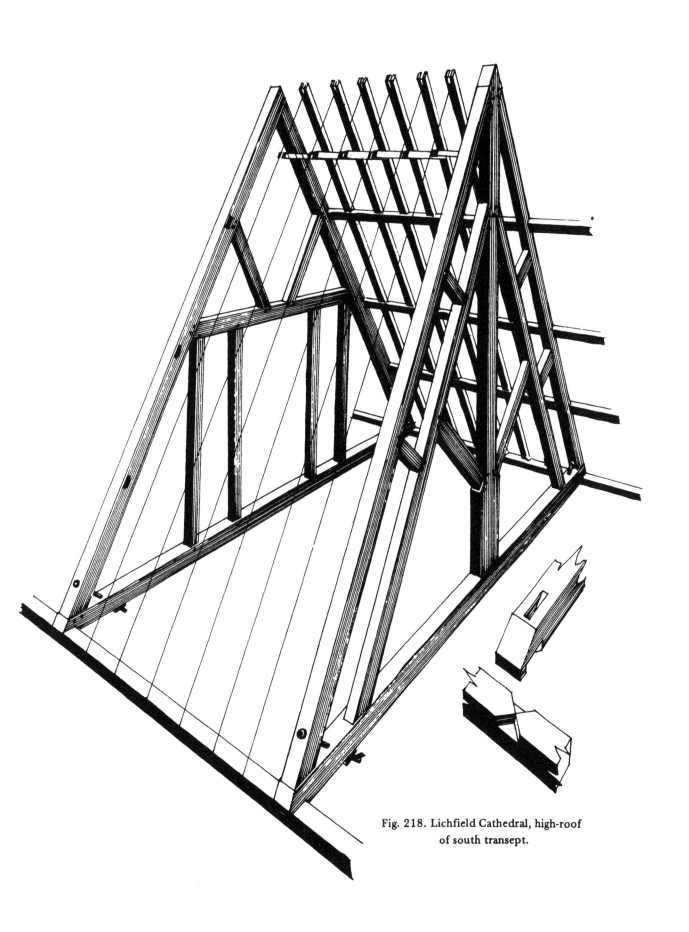

Fig. 218. Lichfield Cathedral, high-roof
of south transept.

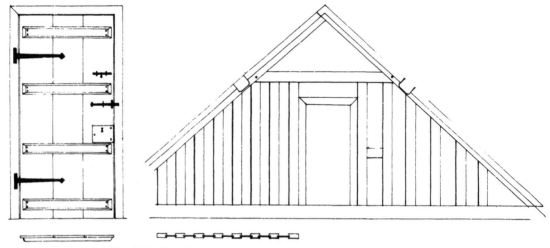

Fig. 219. The George Inn, Southwark, attic partition and door leaf.

The George Inn, Southwark, London

The date of the existing building is fixed by the fire that consumed most of the Borough Market in 1676, after which a Mr. Weyland rebuilt it to the previous plans, and it is his south wing that remains today (W. Kent, 1970, 6). The latter may be assumed to date, at the latest, from 1677. It is an interesting example of soft wood carpentry, built into a three-sided red-brick carcase, of three storeys with attics and a cellar beneath. The galleries facing the former courtyard are jettied, apparently in pine; the roof framing and such other parts of the structure as can be closely examined are certainly of pine. The joists are fitted with diminished-haunch tenons—an important detail in London at so late a date. The cutting of these joints is very exact, and in cases where waney edges necessitated the use of timber of almost quadrant section both the haunches and their sockets were worked to precisely that section. The attics were fitted out as rooms for commercial travellers (then persons of inferior social standing) without great expenditure. One partition wall, typical of those separating the attics, is shown as Fig. 219, which also illustrates the construction method used—thin planks let into the edges of grooved studs. An attic door leaf is also shown, since it is dated, and possesses all its original iron furniture.

The High-Roof of the Eastern Arm, Worcester Cathedral

These roofs, three in number, form a single unit that is well integrated at the transeptal intersection of the main range. The trusses are of oak and have tie beams supported at their centres by foot-strapped and dovetail-headed king posts with two raking struts on each side; side purlins complete the assembly, with double wall plates and, surprisingly, ashlar pieces. The king post irons are forelock-bolted, indicating that at some date during the 17th century major re-roofing took place that apparently eluded the records. The areas between tie beams have an unusual device, namely wedged diagonals in the horizontal plane, which were to ensure the maintenance of right angles. The end of one of these is illustrated in Fig. 220, which also shows a sample of the roof and the foot of a king post. The king posts are also unusual in that they are steadied a little above their feet by longitudinal timbers.

244

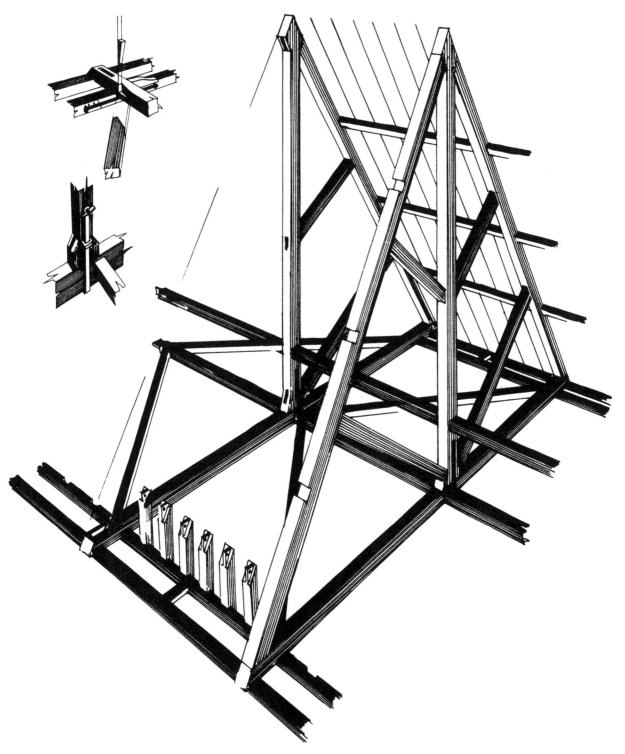

Fig. 220. Worcester Cathedral, high-roof of choir.

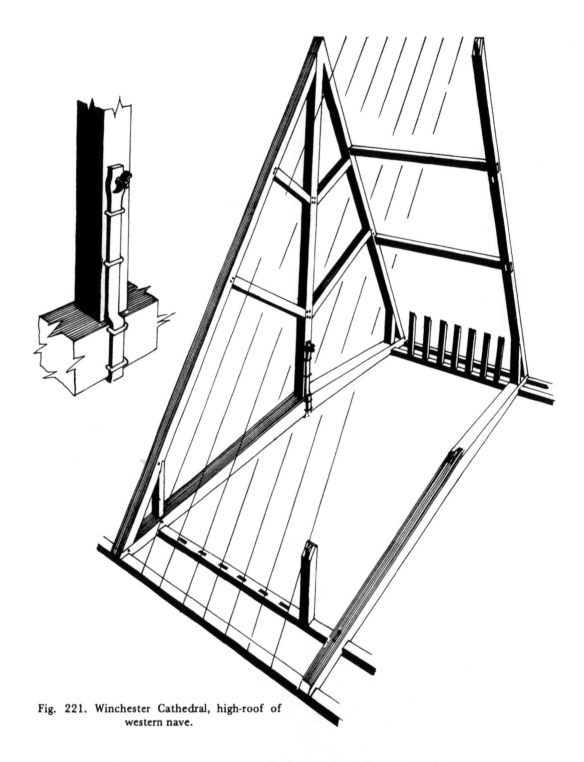

Fig. 221. Winchester Cathedral, high-roof of
western nave.

The High-Roof of the Western Nave, Winchester Cathedral, Hampshire

At the west end the Winchester nave roof has five king-post trusses, one of which is shown in Fig. 221. Each bay has seven common rafters, all fitted with ashlar pieces, and the work was executed in oak. In addition, forelock bolts were used for the foot straps. Documentary evidence as to the construction date of 1699 exists.

The Roof of the Great South Transept, Beverley Minster, Yorkshire

The most readily available record of the repairs effected during the 18th century at Beverley (K. Downes, 1969) gives information on the activities of Nicholas Hawksmoor between 1716 and 1720, but does not mention the south transept, the roof of which is illustrated in Fig. 222. This employed some built tie beams that were still cambered and bolted, the remainder being framed and pegged; some earlier and probably original rafters, bearing notched-lap-joint sockets, were re-used on this occasion. The double wall plating is of interest, as is the framing of the ridge; the bay lengths were short, containing only four common rafters each, and Hawksmoor is considered likely to have been the designer, since this and the roof of the Great North Transept are almost identical.

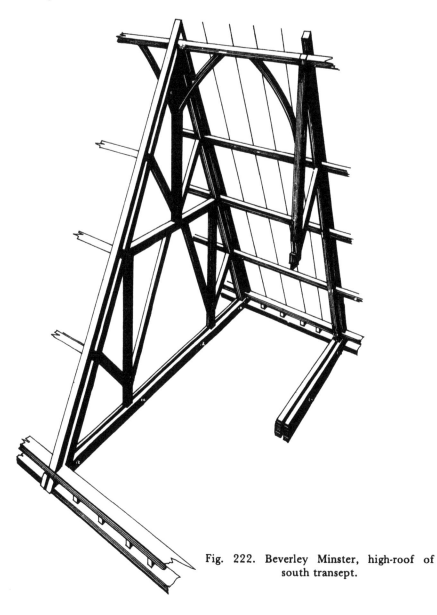

Fig. 222. Beverley Minster, high-roof of south transept.

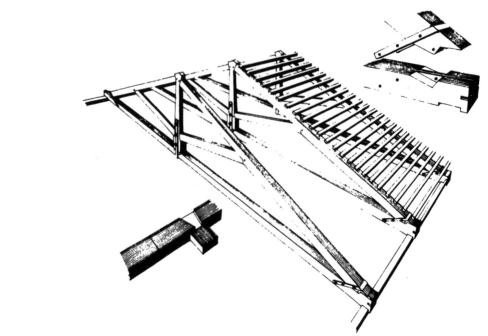

Fig. 223. St. Paul's Cathedral, high-roof of nave.

Fig. 224. St. Paul's Cathedral, high-roof of
western portico.

The High-Roof of the Nave, St. Paul's Cathedral, London

Part of this roof is illustrated in Fig. 223. Its building was completed between 1675 and 1710. The span is unusually wide, being a little more than fifty feet, and this caused great difficulty in finding oaks that were long enough to make the single beams; the quest for them occupied several years. However, the Duke of Newcastle eventually supplied them, and they were delivered on site in 1693. Why unscarfed tie beams were desired is not known, but they certainly were not used for the roof above the western portico (Fig. 224). The latter was an interesting design using shorter timbers and comprising built and cambered beams, with king posts, struts and raking struts. In both roofs Wren used common purlins instead of common rafters, a system that was widespread in the New England States of America.

'G' Warehouse, St. Katharine's by the Tower, Tower Hamlets, London

Until 1972 an enormous warehouse existed here, comprising five long bays and four shorter ones, arranged alternately; the carcase was of red brick with Diocletian windows for the upper floor (originally there was only a ground floor and first floor). This had been built as a brewery tun-house, or vat-house, and the framing for the seating of vats on the first floor had largely survived. The roof, which had *been peg-tiled, was of great size and excellent construction, being queen-posted with raking struts and short secondary principals, with suspended king posts above the collars (Fig. 225). A ridge piece and two side purlins were fitted into each pitch. It seems probable, in view of the unusually complex framing of the tun-floor (Fig. 226), that the architect anticipated that it would have to hold more than the normal dead weight. Porter tuns may have been the reason for this, the drink being in great demand during the 18th century in London, and was heavy to store. A number of structural details are drawn as Figs. 227 to 233.

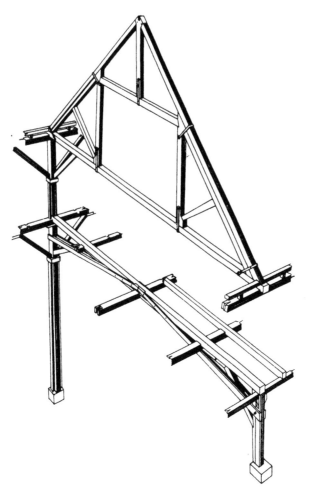

Fig. 225. 'G' warehouse, St. Katharine's, London, frame.

249

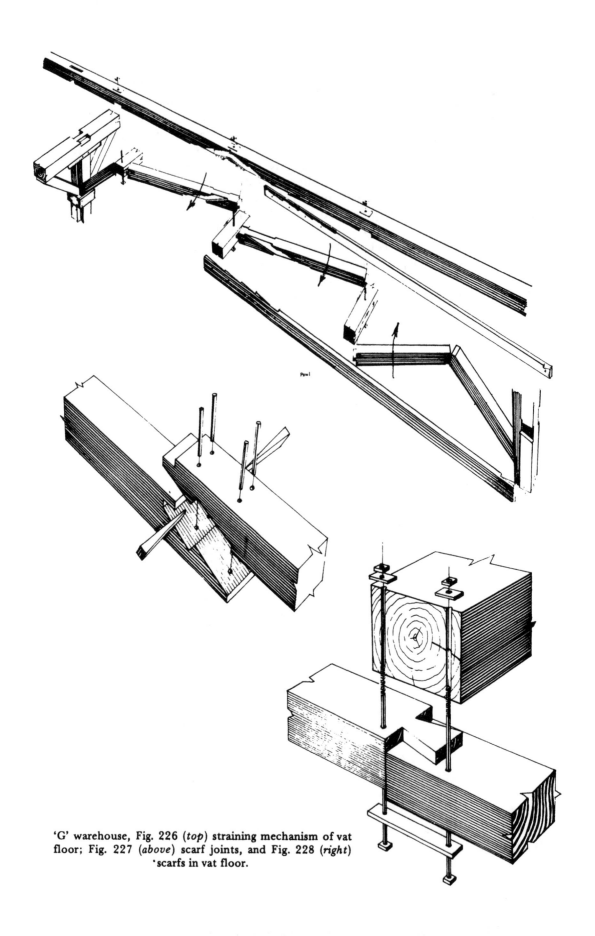

Pawl

'G' warehouse, Fig. 226 (*top*) straining mechanism of vat floor; Fig. 227 (*above*) scarf joints, and Fig. 228 (*right*) 'scarfs in vat floor.

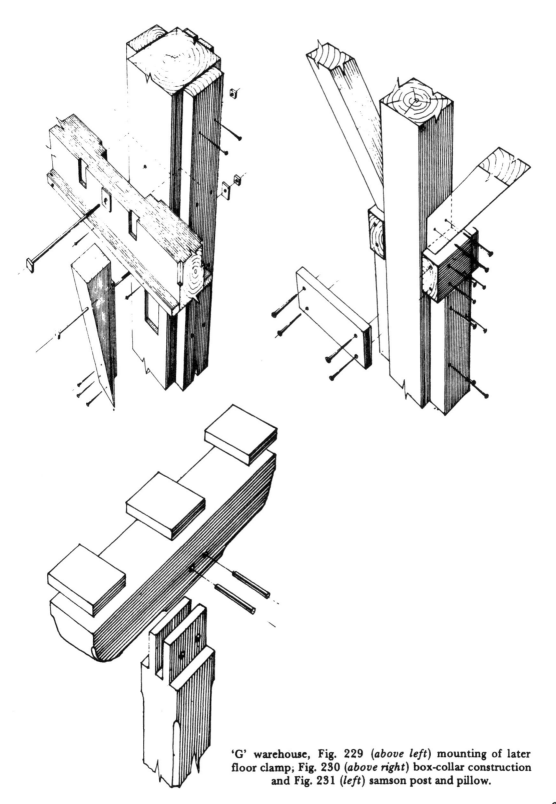

'G' warehouse, Fig. 229 (*above left*) mounting of later floor clamp; Fig. 230 (*above right*) box-collar construction and Fig. 231 (*left*) samson post and pillow.

251

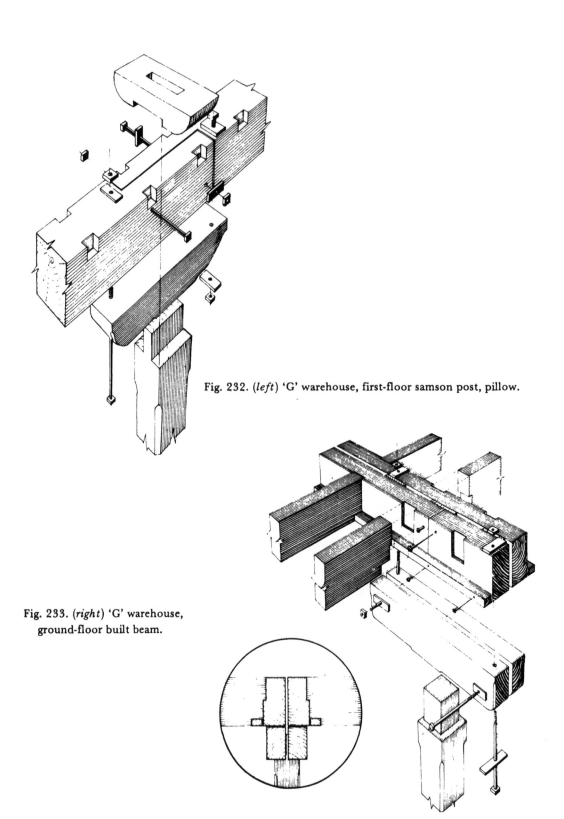

Fig. 232. (*left*) 'G' warehouse, first-floor samson post, pillow.

Fig. 233. (*right*) 'G' warehouse,
ground-floor built beam.

252

The High-Roof of the North-West Transept, Lincoln Cathedral

This roof is of considerable interest, and two of its different trusses are illustrated in Fig. 234, with one iron stirrup enlarged. Screw-threaded bolts were used for these, and the date of the work might be that of its restoration by James Essex in 1762-65. However, the ingenuity displayed here is not apparent in Essex's work at Ely and it is likely that some alternative, if unrecorded, architect was involved. The use of a long bolt from collar to upper collar in the right-hand truss, in order to tighten the whole and distribute a slight pre-stress evenly throughout the triangle, is ingenious, and the same technique has been noted in at least one barn in Essex. Assessed as roofing these designs are good, but if judged as carpentry they are bad, since they are wholly dependent upon their ironwork.

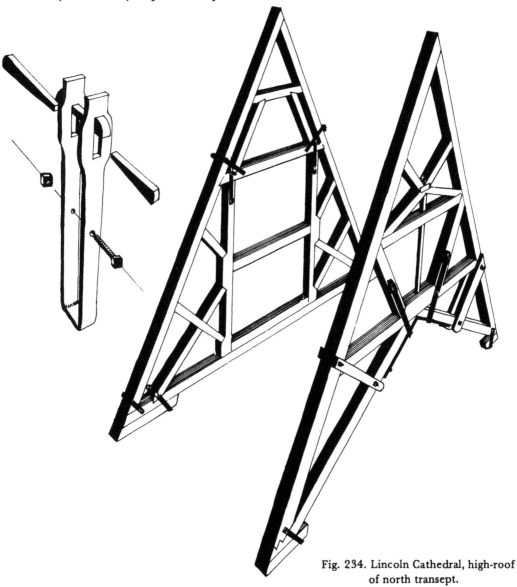

Fig. 234. Lincoln Cathedral, high-roof of north transept.

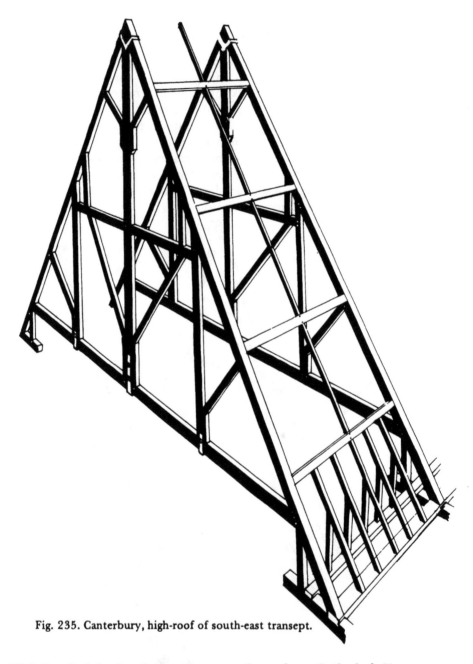

Fig. 235. Canterbury, high-roof of south-east transept.

The High-Roof of the South-East Transept, Canterbury Cathedral, Kent

(Fig. 235.) It is impressive, mainly because of its great size, and appears to be dated by figures cut into a tie beam, possibly by one of the craftsmen working at the time, to 1771. It is framed in oak, which has been spoilt by creosote, and relies mainly upon jointing and pegs with little recourse to ironwork. In principle it comprises king-post and queen-post trusses that are set upon stilted tie beams and are raking-strutted. All the common rafters were framed into the side purlins, and ashlar pieces were used at their feet.

House Mill, Three Mills Lane, Bromley-by Bow, Newham, London

When surveyed in 1973, an almost complete tidal watermill dated by an elaborate wall plaque to 'D.S.B. 1776' survived here, although it had long been out of use as a watermill. It was handsomely carcased in red brick and comprised five storeys, the upper two of which were by definition attics, the eaves level being for bin storage, and the top, as was usual, for sack transportation. A partly sectioned view is shown in perspective in Fig. 236. The floors were massively samson-posted, originally to their full height, but only the upper floors retained their posts when examined. The timbers were mainly of softwood, largely pine, and the reliance placed at this time upon hanging knees and samson posts bears definite affinities with the then current shipwrights' methods of framing. The carpentry, however, was good, and accurately worked; examples of various jointing techniques are illustrated in Figs. 237 to 240.

Fig. 236. House Mill, Bromley-by-Bow, framing and carcase.

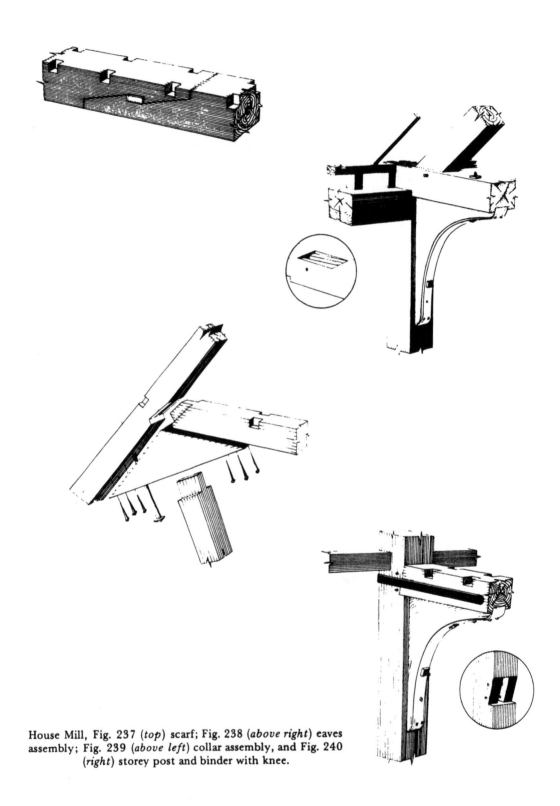

House Mill, Fig. 237 (*top*) scarf; Fig. 238 (*above right*) eaves assembly; Fig. 239 (*above left*) collar assembly, and Fig. 240 (*right*) storey post and binder with knee.

256

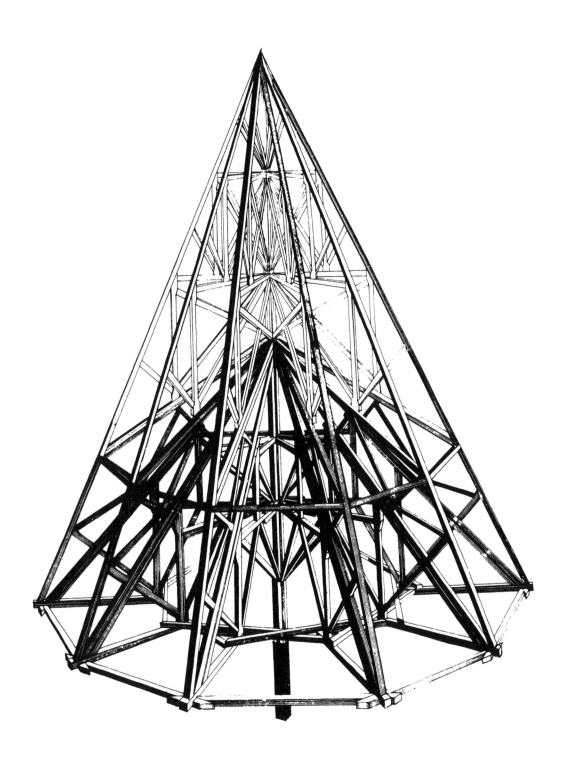

Fig. 241. Lincoln Cathedral, chapter house spire.

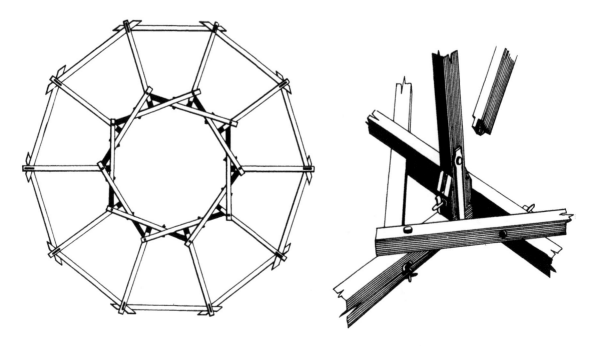

Fig. 242. Lincoln Cathedral, assembly of duodecagon.

The Chapter House Spire, Lincoln Cathedral

The Lincoln chapter house was designed and built under Master Alexander between *c.* 1220 and 1234; it is decagonal, and now has the spire illustrated in Fig. 241, which is the result of two successive building operations. The first new roof was a 10-sided gambrelled cone, resting on a central pier, and tied elaborately with a 'ring beam' (Fig. 242). This roof, it is believed, was designed by James Essex, and the subsequent addition of the spire apex is believed to have been the work of Pearson *c.* 1890. Both works are executed entirely in softwood, and the lower structure relies upon ironwork and forelock bolts, but cannot escape the basic unsuitability of pinewood for such complex carpentry; the timber was splitting badly at the time of inspection. The basis of the design is two queen-post assemblies, one upon the other, set within a polygonal conoid roof built around a spire mast. At the time of raising the apex the arris rafters were doubled and the king-post, raking-strut and queen-post trusses built on, producing a gambrelled appearance; by this means the taller apex was framed.

This must be among the largest and most impressive works of softwood carpentry that have survived, and it possibly derives much of its merit from the inherited and difficult problem of roofing a medieval chapter house. However, despite the limitations that 18th-century English architects felt were imposed upon them by rational king posts and raking struts, it does achieve a remarkable visual quality from within.

258

The High-Roof of the Choir, Sherborne Abbey, Dorset

The choir of Sherborne has a very early and boldly designed fan vault, which had begun to subside seriously by 1856 (J. H. P. Gibb, 1972, 19). The whole fan vault was rebuilt, the operation being completed in three months at a cost of £425. It seems probable that this may also be the date of the high-roof, a part of which is illustrated in Fig. 243. Much use was made of screw-threaded nuts and bolts; the queen posts were paired, as were the collars, and fitted into remarkably elaborate housings on the principal-rafter flanks. The tie beams were 'built' at the eaves, and an example of this is shown to the lower right of the drawing; but no explanation can be offered for this feature, except that the architect was evidently concerned about inflexibility at that point.

Fig. 243. Sherborne Abbey, high-roof.

The Renaissance and after—Summary

Subdivisions of these years into periods of named styles have been made, but in order to summarize the 22 buildings described little more than placing them in chronological order is necessary. In the earliest of the buildings at Staple Inn it is clear that the choice of material and the methods for its use derived from the long medieval tradition; but technologically there was little to be developed, and the major changes were the use of screw-threaded bolts and the importation of resinous softwoods. The continued use at Staple Inn of base crucks is interesting, while the application of the curious standard and pendant decorative carvings to the otherwise medieval design of the Great Hall there reflects a receptive attitude to ideas that had not been previously entertained.

Certain aspects of traditional timber building underwent further development, as is illustrated by the house at Reepham in Norfolk, where the 'hewn' jetty was used, and also the house at Deal Tree Farm in Essex, which was built in a manner which had been established for centuries beforehand. Rural carpenters were prepared to experiment with what were to them new forms of joints, as is shown by the rare use in the Springwell Farm Barn of both halved and bridle-butted scarfs, and bladed ones. It is important in this case that the well-tried joint was used for the main span, and the 'new-fangled' one relegated to the eaves top plate.

Two examples of what was considered to be the best carpentry then possible, wrought without undue regard for costs, were the great new roof for the church of Dore Abbey by John Abel, and the measures taken to reinforce the ancient systems surviving over the Ongar chancel in Essex. These two conform to what was established between 1558 and 1625 as the national architectural style, which is normally subdivided into the Elizabethan and Jacobean. This style had more to recommend it than did mixtures of styles such as those built for Bishop Juxon in his 'new old-fashioned Hall', or the mere curiosities introduced into the hall for Staple Inn.

Of supreme importance was the re-roofing of the north transept of Wells Cathedral in 1661, when imported larch timber, and the screw-threaded nut and bolt were used, both apparently for the first time. It cannot be proved that these bolts were in any way better than medieval forelock bolts, especially for those building uses to which both were put; but the change to the new type, though at first slow, was both inevitable and complete, Wren using them throughout his nave roof at St. Paul's before 1710. This probably resulted rather more from the 'law of supply' than from that of 'supply and demand', because the roof of the Charity School in Limehouse used forelock bolts seven years after St. Paul's adopted screw-threads. Whoever roofed the westward end of the nave at Winchester was conservative in the extreme and adhered to forelock-bolting, as did the designer of the eastern roofs of Worcester Cathedral, which may indicate personal conviction on the part of their designers that the latest was not necessarily the best.

The *George* inn at Southwark is a valuable document, being built of imported softwood before its use at Wells Cathedral, and is of significance because its carpenter persisted in the use of diminished-haunch tenons, knowing that their mechanical efficiency was proportionately retained if applied to softer timber. The tidal watermill at Bromley-by-Bow illustrates a massive structural technique that combined soft- and hardwoods intelligently, together with the use of screw-threaded bolts. This and the Southwark *George* both illustrate the combination of brick carcasing and timber flooring and roofing that had directly resulted from the new fire precautions after the last fire of London. Valid comparisons may be made between the cross-sectional drawing of House Mill and any of the numerous cross-sectional illustrations of men-of-war that have been published (G. S. Laird Clowes, 1932, 78), for the frequent occurrence of transverse beams, hanging knees and samson posts indicates a tradition that was largely common to both shipwrights and carpenters.

The selection of structural assemblies that are illustrated from such buildings show interesting combinations of the available materials and craft techniques, whilst firmly establishing the fact that 'pure' carpentry, such as had been designed for Salisbury's north-east transept roof, had ended.

Appendix One

Scarf joints in chronological succession

Fig. 12 (see p. 19). Scarf with straight bridling, diminished by a splayed edge, and with sallied butment shoulders. This peculiar scarf was used for the upper extension of the Sompting spiremast, and no similar scarfs are known at the time of writing. It is likely to be original, since the cutting of its lower half on the existing and outset lower portion of the timber at some later date than the first building operation is highly improbable. Its sallied shoulders are also inclined, presumably so that the joint might centre itself when compressed.

Fig. 16 (see p. 19). Through-splayed with one under-squinted abutment and one face peg. This was used to scarf the foot of the Sompting spire mast, and it recurs *c.* 1300 on the spire mast at Stifford in Essex (Fig. 111, see p. 125. No peculiar merits can be attributed to this joint, which has apparently survived in both of the cited works because it was subjected to little or no stress, a fact that may imply a complete understanding of the structure on the part of the carpenter.

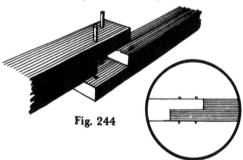

Fig. 244

Fig. 244. Edge-halved scarf with square abutments and two face pegs on its diagonal. This form of the end-to-end joint exists in the barley barn at Cressing, and also in Crepping Hall; in the latter it is unquestionably part of the first period of building and original design, which places it between *c.* 1180 and *c* 1200. This joint lacks integration and needs a supported situation, such as it enjoys at Crepping. Few examples have been found to date, and where found they must be dated by reference to the entire structural context.

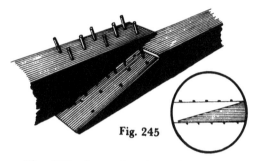

Fig. 245

Fig. 245. Stop-splayed scarf with square under-squinted abutments and 8 face pegs, frequently 10, or 12 face pegs. This is the oldest form of scarf known that seeks to attain some part of the strength of a single timber, and this it can only derive from the large number of pegs that transfix it. The joint is occasionally found to have forelock bolts as well as pegs, as is described in the context of Exeter Cathedral; in earlier examples the bolts may be later additions, subject always to an appraisal of all the evidence. This joint was certainly in use during the 13th century, and at least until the construction of the transeptal roofs at Exeter Cathedral.

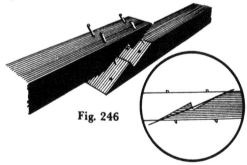

Fig. 246

Fig. 246. Through-splayed and tabled scarf with four face pegs. This form of the joint exists in the bridging-joint of the original Manor House at Little Chesterford, the top plates of Little Hall at Merton College, the side purlins of the nave roof at Chichester Cathedral, the collar purlin of White Roding church, and the bridging-joint of No. 34 The Causeway, Steventon, Berkshire. It was, in addition, used for the top plates of the Court-house at Limpsfield in Surrey and of the service wing of Tiptofts. The pegging varies, as the illustrated examples show; it was a form of the joint designed to resist extension. The period of its use is that of the cited buildings: from *c.* 1190 until the later 13th century.

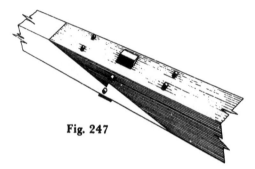

Fig. 247

Fig. 247. Through-splayed and table scarf with through-tenoned tabling, two edge pegs and four face pegs. This joint survives in a top plate of a ruinous barn at Summer's Farm, Doddinghurst, Essex. It is a development of the preceding scarf and dates from the end of the 13th century.

Fig. 79 (see p. 91). An edge-halved and tabled scarf with square and vertical abutments, and transverse key; used for the wall plates of the Salisbury north-eastern transept roof, and for the wall plates of the roof of Winchester Cathedral nave, both of which date from the early 13th century. It is of considerable interest that a variant of this having no tabling, but its key, which was formed by a tie beam's end trenched into both upper and lower halvings, was found to exist in the ruined church of St. Eirene at Istanbul, and dated to a rebuilding shortly after A.D. 740 (Professor C. L. Striker, 1975, letter to author). Where it was used at Salisbury and Winchester this scarf is fully adequate, but its use elsewhere has not as yet been recorded.

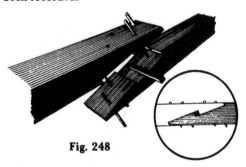

Fig. 248

Fig. 248. A stop-splayed and tabled scarf with under-squinted transverse key, and four face pegs; the joint used for the top plates of the Cressing wheat barn, *c.* 1250. It was also used there for the outshut top plates. This form of the scarf achieved, as intended, a high proportion of the strength of unscarfed timber; two mechanical principles were utilised, the inclined plane and the wedge, and it therefore constitutes a mechanism. This was a scarf capable of resisting every known stress: hog, sag, torque or winding, and flexure. It was designed for 'flying' or unsupported positions.

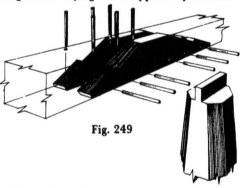

Fig. 249

Fig. 249. A four-part scarf using a pair of stop-splayed and tabled 'fishes', with under-squinted butts (in so far as visible), 12 face pegs and 6 edge pegs. This was the joint used for the arcade top plate for the south aisle of Navestock church, and it is dated by the mouldings to *c.* 1250 (R.C.H.M., 1921).

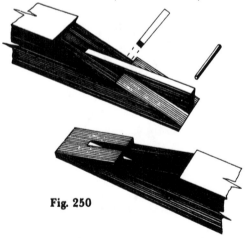

Fig. 250

Fig. 250. Stop-splayed and tabled scarf with under-squinted abutments, tongue-and-groove through the tables, transverse key and terminal edge peg. This development of the preceding scarf was, apparently, incepted when used for the top plates of the Old Deanery at Salisbury during the middle years of the 13th century. It probably offered greater resistance to the winding stresses inherent in a top plate's position. Its period of use is, for the present, confined to the two central decades of the 13th century.

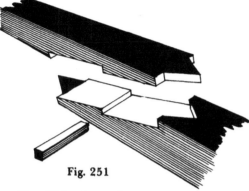

Fig. 251

Fig. 251. Stop-splayed and tabled scarf with inset abutment sallies, under-squinted, with transverse key. This form of the joint was used for the top plates of the larger barn at Sandonbury, Hertfordshire, during the later part of the 13th century under the aegis of St. Paul's, London, who owned the manor. It was a development of the principle, but relied for its enhanced integrity upon greatly increased production costs.

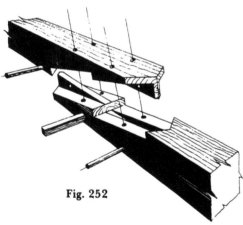

Fig. 252

Fig. 252. A stop-splayed and tabled scarf with sallied and under-squinted abutments, a transverse key, counter tongued-and-grooved tables, four face pegs and two edge pegs. In the present state of knowledge this is the apogee of English scarfing; it survives in the top plates of Place House, Ware, in Hertfordshire. The work has been dated to c. 1295. In

this form the joint-type had reached its perfection, and further elaboration would have weakened it. Subject always to physical tests, this joint should achieve rather more than half the strength of unjointed timber of the same section. The period of its use is apparently the final decade of the 13th century. Comparable examples are known in St. Mary's Hospital at Chichester, St. Etheldreda's church, London, and Wingfield College, Suffolk; but these are simplified, and mechanically inferior.

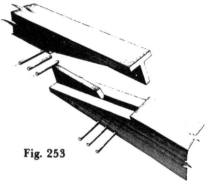

Fig. 253

Fig. 253. Stop-splayed scarf with square and under-squinted abutments, having counter tongued-and-grooved splay, with six edge pegs. This form of the joint survives in the formerly open hall at Wingfield, where it is post-supported and impaled by a tenon that is not illustrated. It was probably used from c. 1300 until c. 1320.

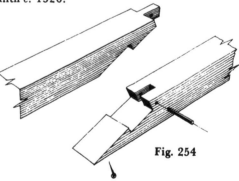

Fig. 254

Fig. 254. Splayed-and-tabled scarf with bridled upper abutment, edge peg, and face spike. This joint was used in the lower framing of the spire scaffold of Salisbury Cathedral soon after c. 1300. The causes of its invention are at the time of writing obscure, and no further examples of its use are known. It was, however, influential, because it combined for the first time the principles of bridling with the tabling and splaying of scarfs.

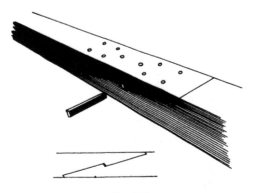

Fig. 255

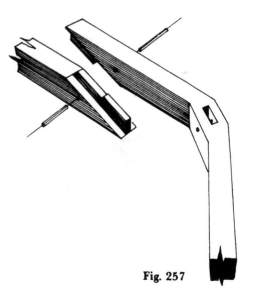

Fig. 257

Fig. 255. Stop-splayed and tabled scarf with square, under-squinted abutments, slip—or feather wedge, and 10 face pegs. This was another development, and one that was highly efficient without greatly increasing cost of production. This joint is known in the collar purlin (which was renewed at the time it was used) of Crepping Hall, the top plates of the barn at Church Hall Farm, Kelvedon, and the top plates of Wynter's Armourie, Magdalene Laver, in Essex (this example is illustrated). The most probable limits for the use of this form would be from *c.* 1290 until *c.* 1325, but this estimate has to be hypothetical.

Fig. 257. A through-splayed scarf, counter tongued-and-grooved, with two edge pegs. This joint was used for the sills of the roof of the chapter house at Wells Cathedral, where it was subject only to compression. The work was completed by 1306, and this form of the scarf may therefore be considered as part of the exploitation of tongued-and-grooved splays that evidently preoccupied master carpenters during the reign of Edward I.

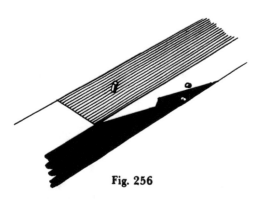

Fig. 256

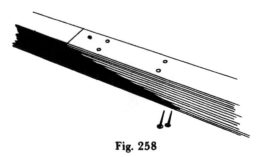

Fig. 258

Fig. 256. A through-splayed and tabled scarf with one face peg and a free tenon between the other tables, edge-pegged. This joint survives in the top plates of the later period of building of Thorley Hall in Hertfordshire, and is datable to the early 14th century.

Fig. 258. A splayed scarf with square, under-squinted abutments and four face pegs, with additional iron spikes. This form of the joint-type seems to have been in use during the third quarter of the 14th century, when numerous questionable practices involving the use of spikes obtained for a short time. This example exists in the collar purlin of one wing of the *Old Sun* inn at Saffron Walden; others are known in Essex.

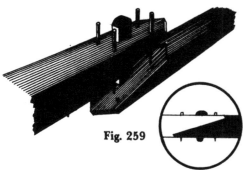

Fig. 259

Fig. 259. Stop-splayed scarf with under-squinted and sallied abutments, face-keyed, with four face pegs. This form of the scarf has little strength, and was normally used in well-supported and stable situations; it occurs in the barn at Great Coxwell, Berkshire, and is frequently transfixed by a tenon instead of a key.

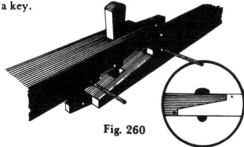

Fig. 260

Fig. 260. Stop-splayed with tapered, bridled abutments, face key and two edge pegs. This form of the joint was used throughout the top plates of the barn at Widdington in Essex, where it may be seen both supported and in 'flying' situations; no failures of the joint have been noted there. No firm date is known for the building, and it is suggested that the second quarter of the 13th century is most probable. Since this and the joint in Fig. 262 are the earliest two at present known to combine bridling with splaying, it is assumed to derive from the joint used in the Salisbury spire scaffold.

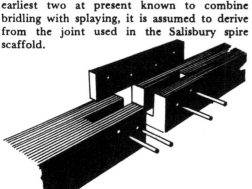

Fig. 261

Fig. 261. Three-part or 'fished' scarf, with square and vertical abutments, free tenon in open-ended face mortises, and four edge pegs. This joint was used for the top plates of the Fressingfield building, and therefore dates from c. 1330; the building is highly complex and structurally unsound, and the scarf has no specific merits.

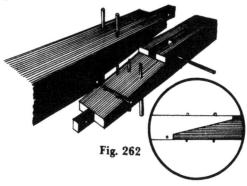

Fig. 262

Fig. 262. The edge-halved and stop-splayed scarf with bridled abutments, four face pegs and two edge pegs. This joint was used for all the plates of St. Clere's Hall, which were carbon-dated to c. 1350. Being a simplified and cheaper version of the joint in Fig. 260, it was probably derived from that source. Examples of this type are quite numerous in East Anglian buildings, and its probable period of use would be from c. 1325 until c. 1400, the latter date indicating use at the vernacular level.

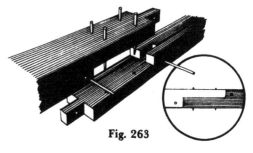

Fig. 263

Fig. 263. The edge-halved scarf with bridled abutments, four face pegs and two edge pegs. This joint is first known in the sill of the London waterfront at Trig Lane, where it was dated archaeologically to c. 1375 (in that example it was grooved to receive a vertical plank on its upper face). Such early examples are normally long in their halvings. The type continued in use, almost without rivals, until c. 1650, the date of the last known specimen in the barn at Rickling Green, Essex.

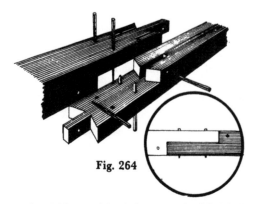

Fig. 264

Fig. 264. An edge-halved scarf with bird's-mouthed and bridled abutments, four face pegs and two edge pegs. This form of the joint was used for barns at Black Notley Hall and Nettes-wellbury, both in Essex. Its period of use was probably the 15th century; it is expensive and looks very well, but has no known mechanical advantages over its plainer form that justify its greater production cost.

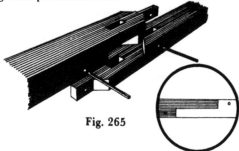

Fig. 265

Fig. 265. An edge-halved scarf with sallied and bridled abutments, and two edge pegs, frequently impaled by a post's tenon. The logical variant of the previous joint, this was used for the outshut top plates of the barn at Netteswellbury, which incorporates both types.

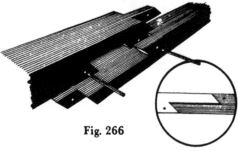

Fig. 266

Fig. 266. An edge-halved scarf with under-squinted bridled abutments and two edge pegs. A good example exists on the rear top plates of the Old Vicarage at Headcorn, Kent.

This form of the joint belongs to the 15th century, and mainly occurred in the earlier part of it. It has clear advantages in positions subject to either hog or sag stresses.

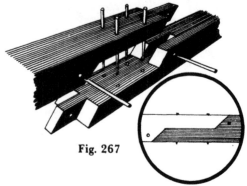

Fig. 267

Fig. 267. An edge-halved and bridled scarf with over-squinted abutments, two edge- and four face pegs; the obvious variant to the last form. This is of the same date range and possesses no known advantages.

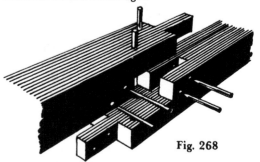

Fig. 268

Fig. 268. An edge-halved scarf with short halvings and long bridlings, all pegged twice along their diagonals. This joint formerly existed in the barn at Brett's Hall, Tendring in Essex (now demolished). Being a contracted version of the root form of the joint it econo-mises on timber. Its period of use was from *c.* 1500 until *c.*1575.

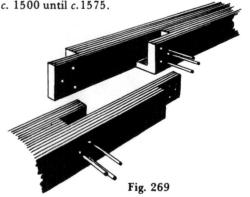

Fig. 269

Fig. 269. A face-halved and bladed scarf with one blade housed, both blades having three edge pegs. An innovation for its times. This form of the scarf is first known in the granary at Rookwood Hall, Abbess Roding in Essex, and later in the barn at Wiggon's Farm, Helions Bumpstead, also in Essex. In the first case the probable date is the second quarter of the 15th century, and the scarf appears there in association with both side-, and centre- or collar purlins. Few examples have been recorded.

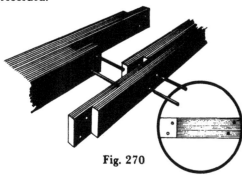

Fig. 270

Fig. 270. The face-halved and bladed scarf with four edge pegs. The earliest dated example known at present is on the top plates of Rooks Hall, Cressing, Essex. This house has a carved date: 1575. This form of the scarf continued in use through the 19th century to the present day.

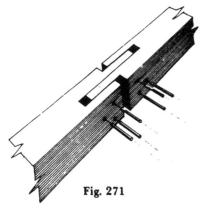

Fig. 271

Fig. 271. The counter-bladed scarf, face-halved with six edge pegs. This is the ultimate contraction of the ultimate form of the joint, and it appears in the house of Clintergate Farm, Redenhall, Norfolk, which dates from *c.* 1590. (It was also used in the Scots Boardman House, at Saugus, Massachusetts, U.S.A., during the 17th century.)

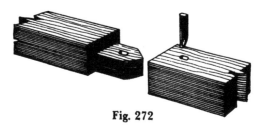

Fig. 272

Fig. 272. The only scarf joint seen by the writer that can be described as 'vernacular'. This exists in the house called Miller's Green Cottage at Willingale in Essex. It has a 'nosed' tongue in a suitable housing and a face key in the form of an eliptically-sectioned peg. The abutments are square and vertical, the whole well cut. It is unlikely that this was thought to possess any peculiar merits, and its origin is totally obscure; if it provides evidence for a regional and idiomatic joint, it is the only example seen in Essex, while the remainder of the house frame is in no way unusual. This joint is, therefore, inexplicable at present.

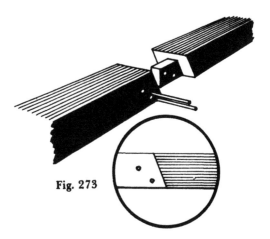

Fig. 273

Fig. 273. A straight bridling of three-quarter depth with squinted abutments and two edge pegs. This joint was used for the original portion of the groundsill of the barley barn at Cressing, which is datable to the 12th century. No other examples of this joint are known at the time of writing.

Fig. 180 (see p. 197). A face-lapped scarf with full-length tongue-and-groove, square vertical abutments and edge pegs of varying numbers. This was used at Canterbury Cathedral in *c.* 1455, and also for the monastic barn at Little Wymondley in Hertfordshire, for which a probable date is 1475 (Radiocarbon,

1967, 489). The same form of joint with two edge pegs, was also used for the top plates of the great barn at Harmondsworth, Middlesex, which can probably be dated to *c.* 1415 (Radiocarbon, 1965, 489–90).

Fig. 274. A straight bridling of three-quarter depth with squinted abutments and over-lipped face, edge-pegged. This was the form of the groundsill scarf that was used almost universally during the medieval period, and it was derived from the earlier example in Fig. 273, which has survived in the Cressing barley barn. This lip was possibly added to avert the entry of water.

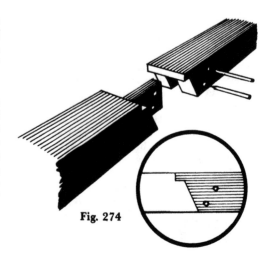

Fig. 274

Scarfing, the General Course of Development

Scarfing evolved from the structural need for timbers longer than could normally be grown; from this need craftsmen created the ideal that the resulting long timber should resemble a single, grown piece, with smooth and continuous surfaces. It was also desirable that such jointed timbers should have a fair proportion of the strength of unjointed ones, and this requirement was responsible for the superlative scarfs of Edward I's reign. The earliest principle used for this purpose in England was the obvious one: the ends of both pieces to be jointed were diminished together.

The Sutton Hoo ship had both stem- and stern timbers scarfed to her keel with stop-splayed joints that were rivetted; her strakes were also scarfed together to produce their required lengths, in contrast to earlier vessels such as the Nydam ship, proving confidence in elaborate jointing. The date was considered to be *c.* A.D. 670 (R. L. S. Bruce-Mitford, 1954, 3–62). This was said to be 'one of the most important historical documents yet found in Europe for the era of the migrations of the Teutonic peoples, in which the settlement of England by the Saxons was an episode' (Bruce-Mitford, 1954, 11). The ship proves the immediate source of the splayed scarfs in England to be European and centuries earlier than the Conquest, by which their subsequent development in this country appears to have been uninfluenced. Two examples of this type survive on the central and original section of the Sompting spiremast, both of which were adapted to resist compression. By *c.* 1200 it seems that terminal and under-squinted abutments had been added to such joints, and their pegging had increased greatly, as shown in Fig. 245. At this stage such scarfs possessed only the strength of their numerous pegs which, until they sheared, made the slight integration afforded by the under-squinted abutments available to the assembly.

A great advance was made when, at some unknown date before *c.* 1250 (Cressing wheat barn), tabling and the transverse key were added to this scarf. It thereby became a mechanism employing two principles, the inclined plane and the wedge, capable of integrating its two parts by the contractile force generated by the driven wedge. This mechanically-efficient scarf was further developed by adding tongues and grooves to its splayed tablings, achieving its perfection in the example shown in Fig. 252 (subject always to further discoveries) during the final decade of the 13th century. The tongues and grooves appear to have originated at Salisbury, where another important feature was also incorporated—the bridled abutment shown in Fig. 254, which was used for the spire scaffold. Variations of these splayed scarfs occurred at Salisbury (C. A. Hewett, 1974, 150) and elsewhere. A good example has been recorded (O. Rackham, W. J. Blair and J. T. Munby, 1978, 110) from the church of the Blackfriars at Gloucester, which has partially diminished face halvings which are counter-splayed, with pegs and iron spikes.

The early introduction of such spikes or forelock bolts into scarfing did not much influence the development of the joint because the competent carpenter required no help from the blacksmith. It is apparent that not only the splayed category of joint was experimented with during this period: less effective joints, such as that used for the barley barn top plates, were also employed, as was the curious type noted for the wall plates of both Salisbury and Winchester Cathedrals (Fig. 93); but the sequence of developments pursued here was the national one, resulting from the highest level of patronage and the masters of carpentry acknowledged by the Crown.

Both the tonguing-and-grooving and the bridling of abutments had, it seems, been incepted at Salisbury as possible improvements to the splayed type of scarf by the middle decade of the 13th century, and these two principles had different effects upon subsequent developments in scarfing. The first enjoyed about 50 years of highly skilled and expensive use that led to the finest scarf joint known in England, at Place House, Ware, in *c.* 1295, and was thereafter found too costly for further use; while the second was to become the principle underlying the succeeding types of the joint that were to be used on a national scale.

The earliest example of the next type is that in Fig. 262, from the top plates of St. Clere's Hall, which was transitional between splayed scarfs and those with halvings. In this form tabling and mechanical integration were omitted, except for the rare example of the type from the barn at Prior's Hall in Widdington (Fig. 260), in which the bridlings were tapered and a face key was used to tighten them before drilling for the pegs. Few examples of these transitional scarfs are known and it seems probable that the next development occurred so soon as to put a period to their economic viability.

By *c.* 1375 the parallel edge-halving with bridled abutments had been adopted, the earliest surviving examples having been excavated at the Trig lane waterfront in London. This type of scarf was to be the one most widely used in England from that date until about 1650, which is the date carved in the barn at Rickling Green (C. A. Hewett, 1969, 161). During the 15th century various refinements were devised such as over- or under-squinted bridlings, and bird's-mouthed or sallied bridlings, both of which assist with date ascriptions, but in its basic form this joint does not assist with the dating of buildings because it was in use for too long a period of time.

Eventually, however, a further striking change took place which rotated the axis of the halving plane through 90 degrees; scarfs became face-halved instead of edge-halved, and their abutments were bladed instead of bridled. The earliest example known at present is that shown in Fig. 269, from the top plate of the granary at Rookwood Hall at Abbess Roding in Essex. This had one blade housed and exists in the context of one of the most important hybrid roof types, which combines both collar- and side purlins; its date is probably *c.* 1425. In this form the new scarf was not, it seems, an immediate success and the next dated example occurs in Rooks Hall at Cressing, again in Essex. It appears that by this time (*c.* 1575) the joint had achieved general usage, and had been simplified by the use of through bladings; this form of the scarf then continued in general use until the present day.

The ultimate development, however, was the contraction of this type into a counter blading that possessed no halvings at all; an example was found in the house at Clintergate Farm in Norfolk, and is datable to *c.* 1590. It is illustrated in Fig. 271, and was also noted in the Scotch Boardman House, Massachusetts, U.S.A.; but there is at this time no evidence for a wide use of the type in England.

Appendix Two

Forms of post-head and tying joints for tie beams

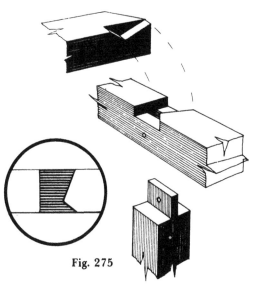

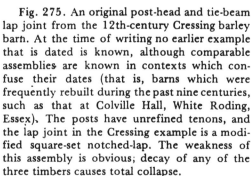

Fig. 275

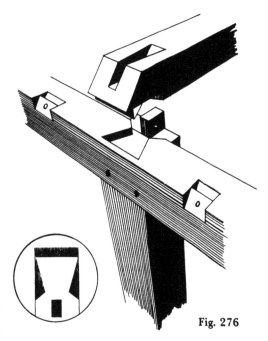

Fig. 276

Fig. 275. An original post-head and tie-beam lap joint from the 12th-century Cressing barley barn. At the time of writing no earlier example that is dated is known, although comparable assemblies are known in contexts which confuse their dates (that is, barns which were frequently rebuilt during the past nine centuries, such as that at Colville Hall, White Roding, Essex). The posts have unrefined tenons, and the lap joint in the Cressing example is a modified square-set notched-lap. The weakness of this assembly is obvious; decay of any of the three timbers causes total collapse.

Fig. 86 (see p. 98). An example from the bell turret at Aythorpe Roding church, c. 1300.

Fig. 276. The assembly used in the wheat barn at Cressing, c. 1250, for both main- and aisle-spans. The post has an unrefined tenon with frontal upstand, the upstand stub-tenoning into the beam's soffit. The lap joint is a lap dovetail with entrant shoulders, the tail being stopped short. This is the earliest *dated* complex of its kind, in which each of the three timbers is jointed securely to the other two. The upstand is the origin of the ensuing jowl.

Fig. 277

Fig. 277. An example from Southchurch Hall in Essex. This has the same type of short and straight-swollen jowl, two unrefined tenons, and a full lap dovetail with over-squinted shoulders. This form was designed to show no cavity when normal shrinkage had taken place. The period of its use was close to *c.* 1300.

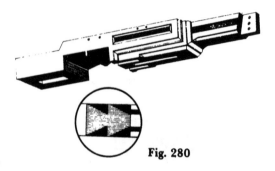

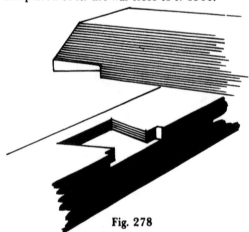

Fig. 278

Fig. 278. Another example of a fully lap-dovetailed assembly, this time with square, housed shoulders. This was also designed to show no cavity after shrinkage or resulting partial withdrawal. The example is taken from the roof of Kersey Priory in Suffolk (C. A. Hewett, 1976, 48-9). It is associated with splayed and tabled scarfing during the close of the 13th century.

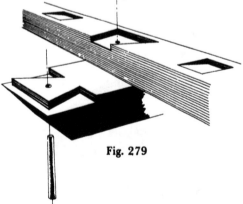

Fig. 279

Fig. 279. The form of lap dovetail used for Lampett's Farmhouse in Essex, which was probably built by Thomas Lampett who possessed the holding from before 1358 until 1411. Similarly acute dovetails were used throughout the 14th and 15th centuries, but the large peg driven through them is peculiarly 14th-century and was an unnecessary practice.

Fig. 280

Fig. 280. The tying joint used for the hammerbeam above the oriel window of the hall of *Staple* inn, London, built between 1580-1. This is a double lap dovetail. Very few other examples are known, and any merits it may have in comparison with alternative dovetail forms would have to be proved by industrial testing; but it is a masterly joint. An example of a similarly doubled but barefaced lap dovetail exists in the Cressing granary.

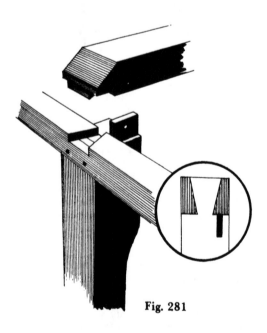

Fig. 281

Fig. 281. The assembly used for the granary at Cressing, dated 1623. This was standard for the 17th century, with a stylishly-cut jowl, full lap dovetail, and the jowl tenon (often named the 'teazle tenon', for unknown reasons) set to one side in order to avoid weakening the 'tail'.

274

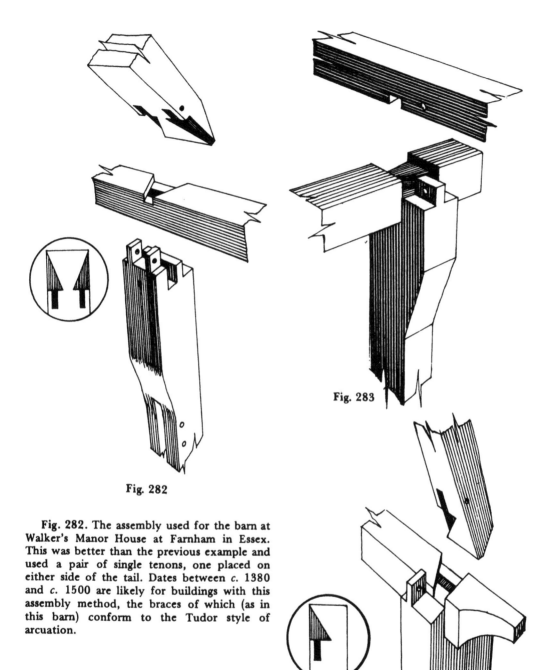

Fig. 282

Fig. 283

Fig. 284

Fig. 282. The assembly used for the barn at Walker's Manor House at Farnham in Essex. This was better than the previous example and used a pair of single tenons, one placed on either side of the tail. Dates between *c.* 1380 and *c.* 1500 are likely for buildings with this assembly method, the braces of which (as in this barn) conform to the Tudor style of arcuation.

Fig. 283. A reversed-assembly from the barn at Prior's Hall, Widdington, in Essex. As illustrated, this eaves assembly postulated the rotation of the jowl of the wall post through 90 degrees, an ingenious device which was not in use in Essex after the early 15th century and which was most commonly used during the 14th century.

Fig. 284. An assembly with jowl and bare-faced lap dovetail, a type that may be found at any date between *c.* 1300 and the end of the 15th century. The belfry at Margaretting in Essex is assembled with this form; but, more importantly, it will be found at every terminal tie beam in buildings otherwise assembled with full dovetails, for the reason that the projecting 'horn' to which the verge boards were affixed was weak and was subjected, therefore, to no lateral strain. In the last instance, of course, it has no dating significance.

Dunmow, Essex; this form was used at the end of the 13th century, and afterwards by eclectic craftsmen. Normal, or entrant-shouldered lap dovetails may be found in association with these.

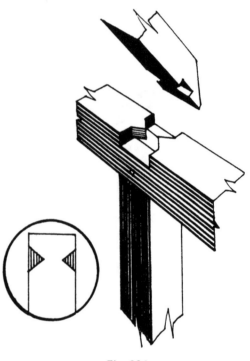

Fig. 286

Fig. 286. The type of assembly employed for the Frindsbury barn, in which the lap-joint used comprised a pair of 'taces' or sallies, aligned across the trench. This was used for the rebuilding of the late-Saxon barn at Belchamp St. Paul's, and the date appears to lie anywhere between *c.* 1200 and *c.* 1400.

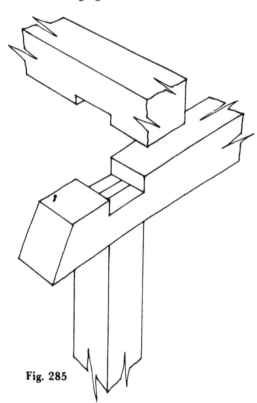

Fig. 285

Fig. 285. A reversed main-span assembly from the barn (now demolished) at Paul's Hall, Belchamp St. Paul's, in Essex. This was carbon-dated to *c.* 1400 (C. A. Hewett, 1969, 111). Comparable examples survive at the rectory, Ashdon, Essex, and in the barn at Stowmarket Museum, in Suffolk. This method was also used for the huge monastic barn at Great Coxswell in Berkshire, where a more complex form of lap joint was employed.

Fig. 94 (see p. 107). On the tall crown posts of the nave roof of Chichester Cathedral are short splayed jowls, similar to those used for the storey posts of Priory Place, Little

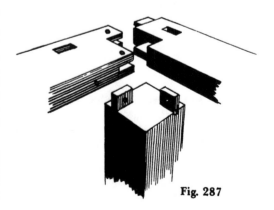

Fig. 287

Fig. 287. An example from the spire at Stock Harvard in Essex, datable to the geometrical era of the Decorated period (*c.* 1250 to *c.* 1300). This method is rare and ingenious and is composed of box tenons through a rebate with inside and outside bridlings, face-pegged. No other examples are known at present.

Fig. 48 (see p. 52). A type of corner joint for top plates that needed a level upper face around the return; this is from the 12th-century bell turret at Bradwell-juxta-Coggeshall.

Fig. 89 (see p. 103). Another solution to the same structural problem, in this case from the wheat barn at Cressing, where it was necessary for the corners of the hipped roof.

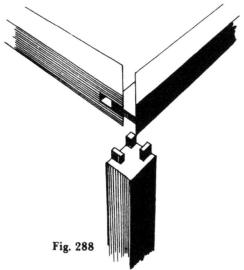

Fig. 288

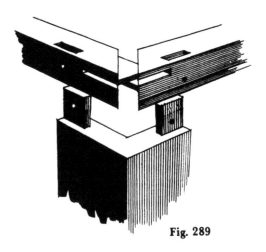

Fig. 289

Fig. 288. A corner joint from the belfry at Blackmore in Essex. This is the mitred bridle with box tenons. Of late-15th-century date.

Fig. 289. An example from the bell turret at Thundersley in Essex which comprises box tenons through a counter-bridled mitre, mounted on a returned corner post.

Appendix Three

Joints used for the framing of floors

Fig. 110 (see p. 124). A joint used for the earliest floors, either ground floors over cellars or first floors in stone or timber carcases; this example is from St. Etheldreda's, in London. Such floors were not framed or jointed, but were 'lodged' in place; their joists were frequently short lengths butted over the bridging-joist. Examples exist at Little Chesterford Manor House (late 12th-century); the tower of Wethersfield church (early 13th-century); and the Old Palace at Croydon in Surrey—to name but three of comparable date.

Fig. 291. Two end-joints with scale of inches. The upper diagram shows a side elevation and an end elevation of a radially set stretcher of the Round Table in Winchester, in which the tenon was cut nearer the soffit than the upper face. The lower diagram shows the same two views of a solar floor joist at Kennington's, at Aveley in Essex, in which the tenon is weakly shallow and only one inch below the central position.

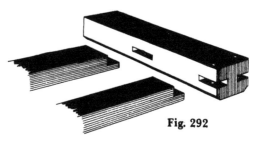

Fig. 292

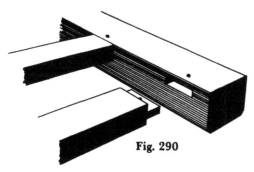

Fig. 290

Fig. 290. The jointing used for the service wing of Tiptoft's Hall in Essex. Central tenons, unrefined; this was a basic joint for construction in general, but in these early and experimental years it was applied to flooring, for which purpose it has no particular merit. In this case the floor was still associated with a samson post, which was placed underneath the jetty. Its date probably falls within the last quarter of the 13th century. This floor was half lodged and half framed; the similarly tenoned and early floor at Priory Place was one-third lodged, two-thirds framed. Both have experimental jetties.

Fig. 292. The barefaced soffit tenon. This was used to frame the floors under the choir stalls of Winchester Cathedral, which were completed by 1309; it was also used for the floors of Baythorne Hall (Fig. 118, see p. 140). An example of this joint was subjected to industrial stress testing, the results of which are quoted on page 284. At Baythorne Hall both the jettying and floor-framing were almost out of their experimental phases, and the samson posts were placed in the side walls. This joint was apparently also used as late as 1398 by the abbott of Battle Abbey, when his Court House was repaired and fitted with a new floor.

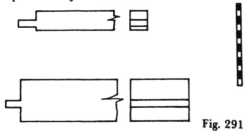

Fig. 291

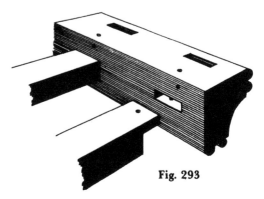

Fig. 293

279

Fig. 293. The barefaced tenon used to support a jetty fascia, a barefaced face tenon; examples may be seen from the pavements of Lavenham in Suffolk. It was used during the 14th century and into the early 15th century.

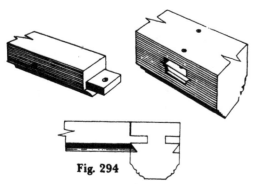

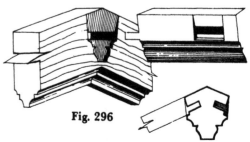

Fig. 296

Fig. 294

Fig. 294. The joist joints used for Tymperleys; central tenons with soffit spurs. This form of jointing is first known in the side purlins of the vicars' houses at Wells, Somerset, which date before 1363 (L. S. Colchester, letter of 1979). Their date at Tymperleys is uncertain, but the moulding used for the binding joists there indicates *c.* 1370; comparable joints exist in the *Boar's Head*, Braintree in Essex (C. A. Hewett, 1969, 199, Fig. 104).

Fig. 296. The end joints used for the library roof at Wells Cathedral, which was completed by 1433. Both the joint uniting the camber beam and the ridge piece, shown at the left of the figure, and that uniting the common rafters and the ridge piece, are identical in principle; only the scribings are dissimilar. Definable as a central tenon of full width, with both face- and soffit shoulders forming one continuous spur bearing, this was evidently developed from the soffit spur, and must have been the point from which Richard Russell derived his diminished haunch.

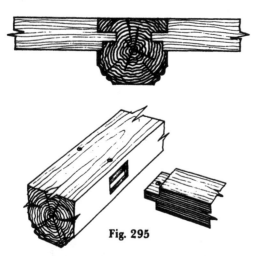

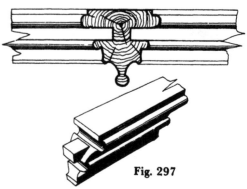

Fig. 297

Fig. 295

Fig. 297. The jointing of the floors in Chichele's Tower, Lambeth Palace, London, dated 1435 by its recorded completion. The mouldings, series of rolls on the chamfer plane, were suggested by the Perpendicular style and prevented carpenters from achieving structural soundness; the tenon was cut on the thick central part of the joist, and the shoulders were scribed.

Fig. 295. The central tenon with housed soffit shoulder, which appears to derive from the soffit-spurred type, which was less easy to cut. It was certainly in use as early as Beaufort's Tower in Winchester, i.e. 1404, and until *c.* 1510. It therefore occurs with a variety of mouldings.

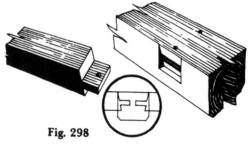

Fig. 298

Fig. 298. The end joint of the common joists in the Jennet Childe's Chantry priest's house at Witham, Essex. This dates from shortly before the completion of the Chantry in 1444, and is a single, and central, tenon mounted upon spurred shoulders; as such it derives from the library roof at Wells and constitutes the joint later adapted by Russell into the diminished haunch.

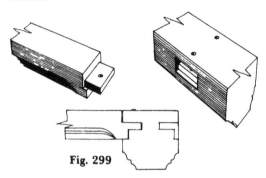

Fig. 299

Fig. 299. Joist-joints from the *Red Lion* inn at Colchester, the date of which is uncertain; however their type was in vogue during the 15th century and until *c*. 1510. It is a central tenon of reduced width with housed soffit shoulder, the mouldings being 'run-out' to form a stop; these were frequently ogees with a return and a hollow. The reduction of tenon width was to avoid cavities in the assembly resulting from timber shrinkage.

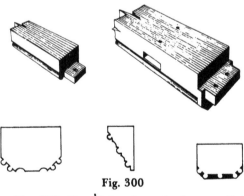

Fig. 300

Fig. 300. Floor joists from the house called Southfields at Dedham, Essex. A building of uncertain dates, with barefaced soffit tenons on roll-moulded common joists, and tenons of reduced widths; the bridging-joists have housed soffit shoulders, as shown. This must mean that the date is fixed by the latest datable features—housed soffit shoulders—to *c*. 1404 or after, and until 1510.

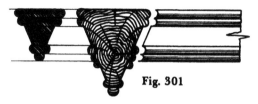

Fig. 301

Fig. 301. The floor joints from Morton's Tower at Lambeth Palace, London, which is dated to *c*. 1495 (Sir N. Pevsner, 1973, 281). This was the ultimate in scribed assemblies; no other form of jointing exists, and the roll mouldings alone support the common joists.

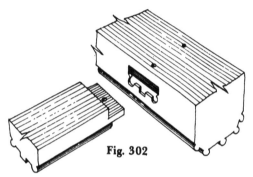

Fig. 302

Fig. 302. The type of scribed joist end used for Jacobe's Hall and Church Farmhouse at Stebbing, in Essex. These had central tenons with housed and scribed soffit shoulders; they avoided the expense of chamfer stops by using the equally expensive scribings, and derive from the plain housings used at Beaufort's Tower in 1404, which continued in use until Paycocke's House, *c*. 1500.

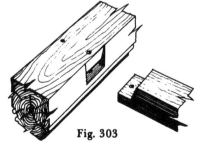

Fig. 303

Fig. 303. The ultimate joist end joint, believed to afford the maximum possible mechanical efficiency in cases where the neutral axis of the main joist is level with the tenon's soffit. This was, apparently, developed by Master Richard Russell, and was used for the side purlins of the roof of King's College Chapel at Cambridge in 1510-12. It is the tenon with diminished haunch, in which the sole object of the tenon is to prevent end-float. Industrial stress tests were conducted on a half-size model of this joint and the observations upon its behaviour are quoted on page 284.

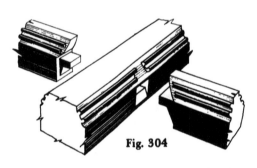

Fig. 304

Fig. 304. Joints from the *Punchbowl* inn, High Easter, Essex. This is a tenon of reduced width with diminished haunch, and is scribed on both major and minor components. This illustrates the frequent conflicts during the Perpendicular period betwixt visual style and structural efficiency; the carpenters could no longer combine both, even at the level of royal patronage. These joints at High Easter have failed under loading, and are propped with various expedients. The date for such work is between *c.* 1512 and *c.* 1570.

Fig. 305. A pair of single tenons, each with diminished haunches, which were used on joists of deep section and are datable to *c.* 1543. This was apparently first used for the Great Standing at Chingford in Essex, which was built for Henry VIII. This was a useful application of Russell's principle to the deep section; it was also used for the ground floor of the Great Hall at the Middle Temple in 1561, from which the illustrated example is taken.

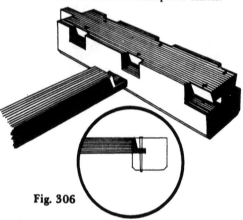

Fig. 306

Fig. 306. An example of work by a rural carpenter who thought that he understood, but evidently did not. It is from the house in Reepham, Norfolk, which was called Candle Court some years ago, when it was visited. This was the tenon with diminished haunch, housed. It is more costly to cut and much weaker in use; since it was found in association with the hewn jetty the date is likely to be 17th-century.

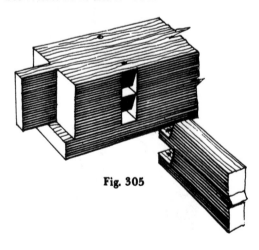

Fig. 305

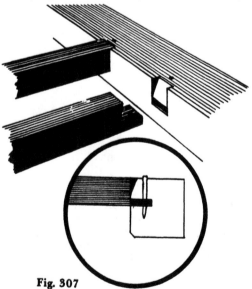

Fig. 307

Fig. 307. A further example of a joint that resulted in deterioration by attempting to improve upon perfection, insofar as the latter had been attained. This exists in the numerous floors of the Old Manor House at Braintree, Essex (C. A. Hewett, 1969, 164). It is a barefaced soffit tenon with spurred face shoulder, and it is resistant to any 'winding' of the joists, but destroys the compression timber of the major joist. It was probably a little cheaper to make, and examples are known during the early 18th century.

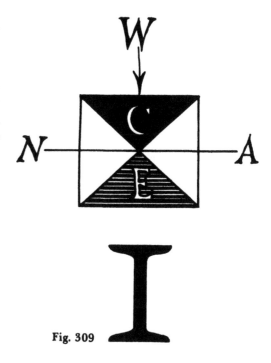

Fig. 309

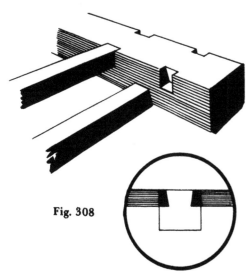

Fig. 308

Fig. 308. Butt coggings, flush with face. This method was employed throughout the late 17th and the 18th centuries in both England and New England. It is without merit, relying solely on the strength of overweight timbers.

Fig. 309. A diagram illustrating the physical behaviour of a floor beam when subjected to loading; both major and minor beams behave in the same manner. The weight is represented by 'W' at the top, and results in the beam dividing along the line 'N.A.', its neutral axis, where there is no stress. The black upper triangle, marked 'C', is then compressed, and the lower, hatched triangle marked 'E' is extended. The black 'I' section beneath illustrates the modern R.S.J., or rolled steel joist, which retains mass at both upper and lower faces. Half-size models of four medieval joist joints were tested to destruction in 1979 by Dr. David Brohn of Bristol Polytechnic's Department of Construction and Environmental Health. The models were made of green oak and had common joists 4ins. square, meeting bridging-joists of 8ins. square; this means that to gain real joint shear measurements (Fig. 310), the cross sections must be multiplied by two, which will multiply the cross-sectional areas by four. It will then become apparent that all medieval floors were stronger than was necessary. The partial rotation of some joists under loading indicates the inaccuracy of the mortising executed by the writer; in normal circumstances this would never occur. Dr. Brohn's report is published here in full, with grateful acknowledgements.

283

TESTING OF MEDIEVAL TIMBER JOINTS FOR C. HEWETT, ESQ.

Date of tests, 17 May 1979

Tests conducted by Dr. D. M. Brohn

Equipment

Denison Transverse Testing Frame, load range 150 kN. Rate of testing, 15 per cent.

Loading Arrangement

The dimensions and arrangement of the support system is shown in Fig. 310(a) and Plate X. The main beam was supported at 10in. centres with the joist placed central relative to these supports. Viewed with the joist in elevation, the left-hand support of the joist is supported on a roller. The distance between the support of the main beam and the end of the joist is 15ins. with the top face of the joint horizontal. The load was applied via a 2in. by 4in. steel plate, the centre of which was 5ins. from the face of the main beam. This arrangement of loading was chosen in order that:

1. The loading on the top of the joist should not interfere with the failure mechanism of the joint, e.g., if the loading plate had been placed close to the face of the main beam, it could have 'clamped' the horizontal shear failure.

2. The main beam was allowed to deflect between the supports in order to be able to identify any tendency for horizontal bending failure of the main beam which did occur in one of the specimens.

The loads given below refer to the jack load at failure. This, however, is not the failure at the plane between the end of the joist and the face of the main beam. That load is the reaction at the main beam support and is a function of the loading position and the span of the beam as shown in Fig. 310(b). That failure load is referred to as the 'Joint shear'. 10 kN = 1 ton.

Test Specimen I (C.13)

    Central tenon and mortice with two face pegs:
        Jack failure 37.8 kN.
        Joint shear 21 kN.

    The mode of failure is shown in the photograph and is a horizontal shear failure at the underside of the tenon, propagating into the joist.

Test Specimen II (C.14)

    Bare-faced soffit-tenon with mortice and two face pegs:
        Failure load 59.5 kN.
        Joint shear, 33 kN.

    This joint had been made so that the angle between the joist and the beam was not a right angle, about 10deg. off, but there is no reason to think that this affected the failure of the joint. The beam failed in tension at the junction of the underside of the joist with the face of the main beam. It would appear that it was a simple tension failure with the tenon acting as a short beam.

Test Specimen III (C.15)

    Tenon with soffit-bearing upper shoulder cut back, two face pegs:
        Failure load 49 kN.
        Joint shear 27kN.

    There was a significant rotation and a pull-out of this joint as the loading proceeded, and the joint eventually failed with rotation between the face of the joint and the main beam. The mode of failure was a horizontal shear failure at the soffit of the tenon portion of the joint.

Test Specimen IV (C.16)

    Bare-faced soffit-tenon with diminished haunch, and two face pegs:
        Failure load 61 kN.
        Joint shear 34 kN.

This joint behaved more favourably than any of the other three in that there was virtually no rotational movement of the joint up to failure. At a jack load of 40 kN there did not appear to be any movement of the joint. Towards the end of the test it appeared that the beam would fail before the joist. There was already substantial splitting along the face of the beam, and one of these splits coincided with the underside of the joist. Close to failure this split opened up and the portion of the beam below the joist was effectively spanning between the supports of the beam. This was deflecting substantially. In the end, however, the tenon broke across the weakest section, diagonally from the face of the main beam to the junction between the haunch and the top face of the tenon.

Dr. D. M. BROHN

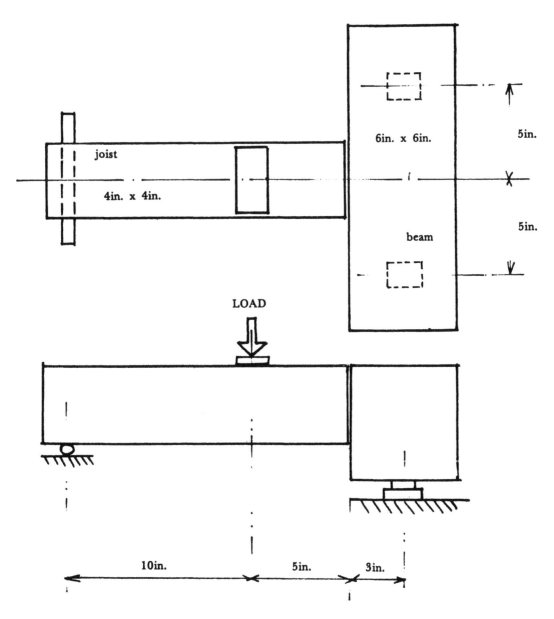

Fig. 310(a). Loading arrangement.

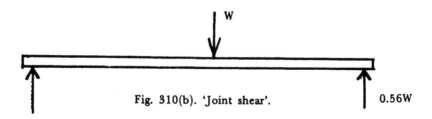

Fig. 310(b). 'Joint shear'.   0.56W

The Development of floor-framing

The few 13th-century first floors described, and the ground floor above St. Etheldreda's under-croft, were lodged; they had no recourse to framing and their components were mostly separate, arranged to rest on the supports provided. Only their floor boards and nails could hold such floors together. It was apparently in order to provide specialised end bays to open halls that such lodged floors were fitted, and in such houses as the Bury at Clavering an attic was created in this manner over the service end. The next available evidence is provided by the Tiptoft's Hall service wing in Fig. 112, in which a jetty was contrived by tenoning two short side-girt extensions to the storey posts. Not far distant, at Priory Place in Little Dunmow, another curiously experimental set of jetties was built, as shown in Fig. 115; these used twice as much timber as was normally used in later times, and achieved less projection as well as less security. A study of Fig. 115 will show that two side girts were employed, one for the ground-storey wall and the other for the first-storey wall and transverse sill; every other joist was supported by an arched brace. In both of these experimental jetties the only joint used to frame the floors was the unrefined mortise and tenon, as shown in Fig. 290.

It seems unlikely in the light of the evidence for the development of almost every category of joint, or carpentered item, that the single and central tenon—such as was used for the floors of Priory Place, or Tiptofts service wing in the 13th century—should have been moved to its soffit-tenon position, c. 1307, as a single refinement. This would constitute a more marked stage of development than normally occurred. There are two buildings among those described, Stanton's Hall and Kennington's, that have framed service-end floors which establish a transitional phase; their joists have single tenons of the full width which were cut an inch nearer to the soffit, or an inch beneath central (Fig. 291). The vast majority of dates are open to question (including those suggested), and in view of the fact that this same peculiarity was noted when examining the Round Table in Winchester Castle Hall, a case for it is established. This suggests that late in the 13th century the joist tenon was deliberately moved *towards* the joist's soffit, and later became barefaced and was positioned off the soffit. The dating of this event must, however, be hypothetical; but if the Round Table is ultimately proved to date from the reign of Henry III (C. A. Hewett, B.B.C. T.V., 1977), then the gradual improvement of the tenon as a horizontal bearing-joint would perhaps have begun a little before 1250, and continued in the off-centre form until the barefaced soffit tenon was incepted a little before 1307 at Winchester Cathedral. Were this supposition correct, the dates of both Kennington's and Stanton's Hall might have to be modified, and both ascribed to before c. 1300. There is no known evidence to conflict with such an ascription, since all that is known of Kennington's is that Sir Henry Gurnett died possessed of it in 1345, at which time he rented the house from Prittlewell Priory. The pronounced scroll mouldings of Kennington's are its strongest dating evidence, and according to Forrester their maximum popularity was between c. 1280 and c. 1340. As to Stanton's, the documentation establishes only the fact that a messuage of that name existed in 1306.

The first competently-framed set of jetties among the examples are those of Baythorne Hall (Fig. 124). There is no known date for the design of Baythorne Hall, but in view of the markedly water-holding base of its main crown post and the use of barefaced soffit tenons for the floors in both of the cross wings, a date shortly before 1300 could be defended. In this case samson posts were used in what was evidently their final role: as additional wall-frame posts, placed there solely to support the bridging-joists, and thereby the floor. The method used at the Tiptoft's service wing was similar in that the samson post was also placed in a wall, beneath the jetty. It appears, therefore, that central unrefined tenons were used until c. 1250, followed by tenons set beneath centre by about one inch, until barefaced soffit tenons were introduced possibly a little before 1300. The latter seem to have been employed until close to the end of the 14th century, if the tenuous evidence of the Limpsfield Old Court Cottage is admitted. It is clear that a 14th-century first floor was fitted into the building, and an occasion is certainly documented (C.P.R., Vol. VI, Oct. 1398) when such damage as might have occasioned this change could well have been done. If this is the case, the abbot of Battle was using the joint as late as the autumn of 1398.

The next development of the joint concerned—the mortise and tenon—appears in the first floor of Beaufort's Tower at St. Cross Hospital in Winchester, which is dated to 1404 according to that institution's records. This was the single, and central, tenon with housed soffit shoulder (Fig. 295), a form of the joint for floor framing purposes which was certainly in use until c. 1495 when Paycocke's House was built, a period of about ninety years. Before this long use of house soffit shoulders had begun, an alternative had been introduced in the roofs of the Wells vicars' houses, which were completed by c. 1363 (L. S. Colchester, 1979, letter to Author): this was the central tenon with soffit spur illustrated in Fig. 154.

It is evident that developments in roof framing would affect floor framing, because the loading of roof pitches or planes could well be equally severe, frequently combining the weights of both lead and heavy snowfalls. At the time of writing the earliest example of a floor with soffit spurs at the ends of its joists is that illustrated in Fig. 294, showing Tymperleys in Colchester. This house is not dated, but for the fact that Dr. Gilberd is known to have been born in it, which indicates a date prior to 1543 (Gilberd was physician to Elizabeth I). As shown in the drawing, this joint is there associated with the double-ogee profile with a roll set between the ogees; furthermore the ogees are of an unequally emphasised contour. This is equally true of the profiles worked upon the roof timbers of the Brothers' Hall at St. Cross Hospital, dated by the records to 1372, and a similar date is possible for the Tymperleys floor. Within a short time a further development of the soffit spur appeared in the roof of the Wells Cathedral library, built under Bishop Bubwith and completed in the relevant (northern) bays by c. 1433 (L. S. Colchester, 1979, 15). This was applied at Wells to the ridge pieces of the library ceiling, and it took the form of a single tenon mounted on a spurred bearing (Fig. 296). Evidence that this roofing development affected flooring is provided by the priest's house build for the Childe's Chantry priest at Witham, illustrated in Fig. 198. The date of the latter is necessarily very close to that of the chantry itself (1444).

The good timber craftsman now had two equally viable alternatives, and these two joints continued in use, with some variations, until the ultimate and most efficient form displaced all others. During the 15th century the Perpendicular style seriously affected structural designs, as exemplified by the floors of the two towers built at Lambeth Palace, by Bishop Chichele in 1432, and Cardinal Morton in 1495. This resulted in a preference, based purely upon visual appearances, for series of roll mouldings worked on the chamfer planes of ceiling joists, producing knife-edged joists laid with their sharp edges downward. These were very weak for load-bearing purposes and almost impossible to joint together at right angles; scribed forms of unrefined tenons (Chichele's Tower) and ultimately scribed abutments (Morton's Tower) were used. The better craftsmen combined the housed soffit shoulder with the new style in mouldings and used the scribing to fit the shoulder into the major joist. This is illustrated in Fig. 302: its date range seems to cover the reign of Elizabeth I, but no positive evidence is available.

Apparently Richard Russell, master carpenter to the king, incepted the ultimate joint for this and similar purposes. The work was the completion of King's College Chapel at Cambridge, and the joint was used for the side purlins of its roof (Fig. 194), between 1510 and 1512. It is clear, in the light of recent researches, that this was a consecutive development of the spurred shoulder in which the tenon again returned to the soffit plane, becoming a barefaced soffit tenon with spurred shoulder (now defined as a diminished haunch). Tests have proved that this was of the maximum possible efficiency, and it was in use without further modification as late as 1697, when the new *George* inn was built at Southwark; but alternatives were used (perhaps soon) thereafter, and the dates can rarely be ascertained. Some alternatives, such as that at Reepham (Fig. 306), were both inefficient and more expensive, and the subsequent succession that ended in the butt cogging with flush face was retrogressive.

Appendix Four

Lap joints for secondary purposes (such as bracing)

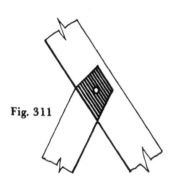

Fig. 311

Fig. 313. One of the disused lap joints from the belfry at West Bergholt in Essex. According to archaeological evidence a possible date for these timbers is *c.* 1000. One can see that they were designed to resist longitudinal withdrawal, because they were worked on adjacent faces of corner posts for a tower structure, probably standing against the west end of the church. Their significance lies in the obtuse angularity of the notch, cut in a form able to offer minimal physical objection to withdrawal.

Fig. 311. The form of lap joint used for the wall anchors of the Sompting Rhenish helm are among the earliest examples known to date. These are also illustrated in Fig. 17. These joints were withdrawable, and were designed only to resist shearing stresses exerted in the plane of the roughly triangular figure which the assembly produced. Their date is uncertain, but must lie between *c.* 950 and *c.* 1050.

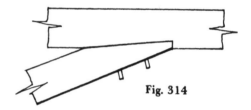

Fig. 314

Fig. 314. A different type of lapped joint, and one which is also found in other European countries. Its date range ends early in the 13th century. The example shows the rafter couples at Chipping Ongar (datable to *c.* 1075).

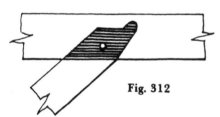

Fig. 312

Fig. 312. Another form of lap joint found in a re-used context, upon one joist of the floor in the Sompting tower illustrated in Figs. 18 and 26. This diagram shows the joint as a plan view; both its date and its original purpose are uncertain.

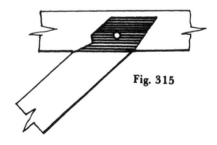

Fig. 315

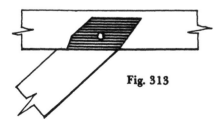

Fig. 313

Fig. 315. A notched-lap-joint from the barn of Coggeshall Abbey, dating from between 1120 and 1147, for which a carbon date of 1130 ± 60 has been derived. This is the earliest example known to have a refined angle of entry, offering maximum objection to longitudinal withdrawal.

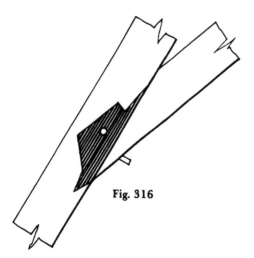

Fig. 316

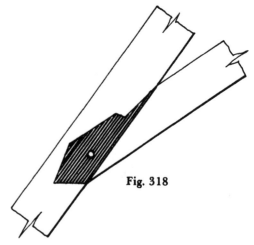

Fig. 318

Fig. 316. An open notched-lap-joint from the lean-to roof of the south triforium to the quire of Wells Cathedral. This has a refined entry combined with an extended 'tip' of uncertain purpose; visually it resembles the socket existing on the re-used joist of the Sompting tower floor (Fig. 312) and being situated in the eastern arm of the church—a part which was built between ?1176 and 1184—it may date from *c.* 1180.

Fig. 318. Joint from the main-span roof at Wells Cathedral, which dates from before the Inderdict of 1209. This has the 'open' notch with refined entry angle.

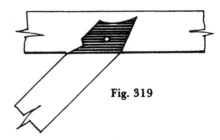

Fig. 319

Fig. 319. Joint from the Bury at Clavering in Essex. Of refined entry with open notch and manneristic curvature to its upper end. Of uncertain date.

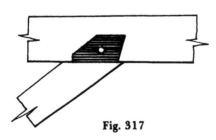

Fig. 317

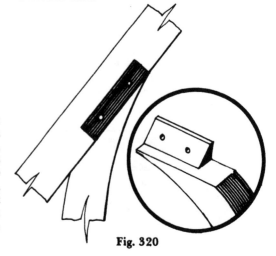

Fig. 320

Fig. 317. Joint from the foot of one of the passing braces of the belfry at Navestock in Essex, carbon-dated to 1180±60 (Radiocarbon, 1964, 338). This is of 'archaic' profile and has the unique feature of a spur at its base, for no recognisable purpose. As shown in Plate XII, the Navestock joints did, in some cases, split due to the inappropriate profile used; an occurrence which illustrates the incentive to design stronger joint forms.

Fig. 320. The type used for the roof of the Greyfriars' Church at Lincoln, and for the roof of the north transept of Tewkesbury Abbey: its date range is apparently from *c.* 1250 until *c.* 1350. This type is difficult to define.

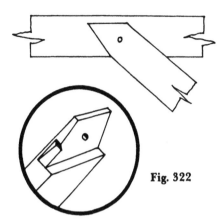

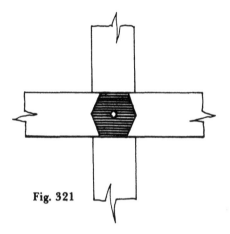

Fig. 321

Fig. 321. From the south transeptal roof of Exeter Cathedral, which is firmly dated to 1290. This can be defined as a counter-sallied cross halving and is attributable to the medieval architect Thomas Witney, who is proven to be a master who was both mason and carpenter, in addition to being an outstanding and advanced designer (Dr. J. H. Harvey, quoted in Hewett, 1974, 25).

Fig. 322

Fig. 322. This joint is from the nave roof of Wells after the Interdict (i.e., *c.* 1213 onwards). It is the secret notched-lap-joint, with refined angle of entry. This form was stronger because the surface flange of timber attached the objecting timber mass by its face, as well as by its edge.

Fig. 46 (see p. 51). A purely compressive lap joint, used for the ashlar pieces of the roof of Waltham Abbey, *c.* 1130. This example contrasts sharply with others, such as the ashlar pieces originally present in the nave roof of Wells, in which unwithdrawable joint forms were apparently used in compressive situations where they were unnecessary.

The use of lap jointing

Lap jointing, being a technique that does not require any highly-developed carpenters' tools, seems from our current viewpoint to have derived from archaism; but apart from the threshold- or corner-joints of Scandinavian log buildings, which were presumably axe-wrought, there is no evidence for this assumption. Its use derived from a desire to achieve long and continuous timbers that passed through buildings or other frame systems in one piece, and supplied all the strength and integrity that was possible within an elaborately-framed assembly. It seems that the earliest known examples, such as on the Sompting wall anchors, were designed to resist any but stretching stresses, the form of stress to which this category of joints was eventually to become peculiarly adapted.

The earliest known examples that were notched to resist withdrawal seem to be those illustrated in Plate I and Fig. 313, which exist upon the re-used timbers in the belfry at West Bergholt. In these the objection to withdrawal is the least possible, amounting to little more than a reduction of the width. In Fig. 317, the type used for the timber tower at Navestock, the objection is greater, the piece removed from the lap amounting to a right angle. This is the 'archaic' profile, which is illustrated in Plate XII, and was structurally weak in use. It had been realised by *c.* 1130–47, when the barn was built for Coggeshall Abbey, that a more subtle entry angle such as is shown in Fig. 315 was less likely to fail; and open notched laps of slightly differing profiles were widely used, all with refined entry angles. This type was used for the roofs of Wells Cathedral up to 1209 (the Interdict), when building ceased until 1213, when it was resumed, and the 'secret' notched-lap-joint appeared in otherwise identical roof couples. As was realised by Dr. Rackham, this form of the joint was stronger because the small volume of the timber piece that resisted withdrawal was now attached from two faces, or planes, instead of just one (O. Rackham, W. J. Blair, and J. T. Munby, 1978, 117). In these most refined of notched-lap-joints the thickness of the lap decreased towards its end, as is shown in Fig. 322. In this highly-perfected form the weakening of the matrix timber is reduced to the essential minimum, and the most perfectly cut examples of it seen by the writer exist in the scissors'-foot joints in the eastern apsidal roof of Westminster Abbey (Fig. 100). This reduction of lap thickness is present in the Cressing wheat barn of *c.* 1250, and appears to have been introduced between truss number 7 and number 8, counting west from the crossing tower of Wells Cathedral, in the North Nave triforium—i.e. between 1200 and 1213.

Evidently there was a clearly discernible development of unwithdrawable lap joints during the Early English period. Examples of joint failures such as the collapse of the archaic notch at Navestock in Plate XII illustrate the necessity of continual effort to produce carpenters' joints of the greatest possible mechanical efficiency. In all categories efficiency was ultimately achieved, but the results were not simultaneous because different types of joints were developed as a result of the varying needs of their times; for example, scribed abutments evolved during the Perpendicular period. Only after various forms of any given joint had been developed was it possible for a carpenter to select which one best suited this purpose, a fact which is well illustrated by the roof couples surviving in the Romsey Abbey precinct.

Fig. 321 at present seems peculiar to the southern transeptal roof at Exeter Cathedral. It is an interesting assembly, but one that does not seem to recognise the normal immovability of cross halvings, the shoulders of which effectually prevent any movements. The compressive form of joint shown in Fig. 313 represents the oldest couples at Chipping Ongar; it does not properly belong in the category of lap joints, but neither does it appear to justify a category of its own, having no variants that are known at the present time. This form was noted by Deneux in France, and by R. Reuter in Germany, with early dates in all cases (Hewett, 1969, 98). It has, in fact, been recorded by Reuter that the terminal abutments are not square. This feature allowed the secondary timbers to be driven sidewise into the assemblies, thus increasing their straining action in the process; finally they were pegged in the face plane.

The type of lap joint shown in Fig. 320 exists in two important roofs, Lincoln's Greyfriars' and Tewkesbury Abbey's northern transept. In the latter their use was obviously eclectic, because perfectly normal lap dovetails were used in the same frames. No such joints have been seen dismantled, but it is probable that the thickness of the lapped portion is greatly increased towards its shoulder, for without such provision the joints would have fractured (a hypothetical view is drawn inset). The manneristic open notched-lap shown in Fig. 319 typifies those at Claveringbury, which is itself of uncertain date; examples also occur in the Cressing barley barn. A preoccupation with lap joints having numerous concavities is much in evidence in Hesse, West Germany, where elaborate profiles were used instead of notches to resist withdrawal; it seems that plain notches only occur there in earlier works such as the apse roof at Grossenbüseck (Hewett, 1969, 97).

Early relationships between wall framing and first-floor mounting

The diagrams constituting Fig. 323 represent end walls of houses all of which today conform to the H-plan; frequently this has occurred by growth.

At 'a' is a longitudinal section of the stone-built service wing of Great Chesterford Manor House. This has a lodged floor on a longitudinal bridging-joist, which is itself mounted upon samson posts with bolsters and braces. The date of this example is not known, although *c.* 1200 has been suggested, but it appears to be earlier in the light of contemporary studies. Comparable floors are known to exist at St. Etheldreda's, Ely Place, London; Nos. 39–43 The Causeway, Steventon, Berkshire; the tower of St. Mary Magdalene and St. Mary the Virgin at Wethersfield in Essex; underneath the Guard Room of the Old Palace at Croydon in Surrey; and in the building known as King John's Hunting Lodge at Romsey in Hampshire. On a typological assessment all these should date before examples with samson posts placed in the line of their walls. The Steventon example provides a rare case of this type of first floor existing inside a timber-framed range, when its effect was to necessitate the use of side girts for the support of the common joists' ends. The majority exist in rubble or masonry carcases, and require an off-set on the inner face of the walls for the same purpose.

At 'c' is a similar longitudinal section of the service wing of Tiptoft's Hall in Essex. In this case the floor was adapted for the production of an end jetty, and the oversailing common joists were tenoned into the last of the transverse joists, which appear to have been lodged. A samson post was placed almost centrally under the jetty in the ground-floor end wall, which was not fitted with any studs. No comparable examples are known.

At 'b' is a similar section of the service wing of Baythorne Hall at Birdbrook in Essex, which can be dated both by comparison with the choir stalls of Winchester Cathedral and with the pronouncedly water-holding base of its crown post to some time between *c.* 1250 and 1309; the latter is the date of the Winchester stalls, and the former marks the start of the decline in popularity of water-holding bases. In this case the first floor was framed and mounted upon four samson posts placed in the side walls, the carcase frame thereby being relieved of the weight of the first floor. A comparable system survives in the base of the tower of the church of St. Lawrence at Upminster in Essex, which has samson posts placed along the northern and southern walls, transoms across, and lodged floor joists. It was ascribed to the 13th century (R.C.H.M., 1921, 160).

At 'd' is a similar section of the service wing of Priory Place, Little Dunmow, in Essex; here the jetties are experimentally framed. It is of uncertain date, but typologically it should pre-date the Baythorne Hall example, and a date within the last quarter of the 13th century is proposed for it. The diagram at 'e' represents the solar wing of Winter's Armourie at Magdalene Laver, Essex, which has an early example of flooring that is both framed and jettied and also integrated with the storey posts. It is in such early examples that the storey posts were severely weakened by excessive mortising at first-floor level, where two side-girt- and one binding-joist mortise met in a common cavity at the post's centre. Such over-mortised posts are to be found in broken or repaired conditions in a high proportion of early houses and barns, the Grange Barn at Coggeshall providing a very good example with many main posts broken at the arch braces' springing-levels. As is apparent from the illustrated selection of buildings, the framing of timber walls involved a series of different methods.

During the 13th century studs of squared timber were employed together with infill material, and to support cladding; such early studs were always widely spaced, as is shown in the drawings of the Cressing wheat barn, Coggeshall Grange Barn or Priory Place. From about 1300 onward (as can be seen in the drawing of Baythorne Hall) studs were more closely set, and were involved in the development of wind bracing, to accommodate which they were trenched obliquely across their external faces. It was during the Decorated period that such braces were fitted on the most lavish scale, as shown by the Fressingfield 'stables'. Previously, as at Priory Place, wind bracing had not been fitted. Stability was achieved by the continued use of long passing braces, but the vogue for decorative braces was there apparent in the form of saltires placed on either side of the king studs in the gables.

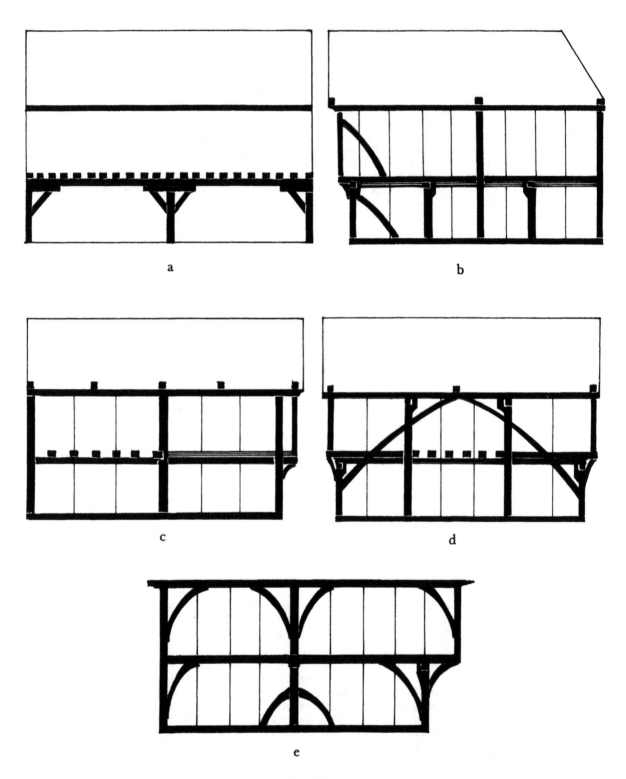

a

b

c

d

e

Fig. 323

294

Studs were multiplied and set more closely together throughout the 14th and 15th centuries, in such areas as could afford them, and soon after the Dissolution of the Monasteries they attained their heaviest cross sections simultaneously with their closest spacing. This was undoubtedly due to the rash and immediate exhaustion of the timber from monastic woodlands, which was expended by the new owners upon ostentatious houses for themselves, and in these cases the interstices between the studs were frequently narrower than were the studs themselves.

During the 16th century a non-functional fashion in wind braces appeared; it seems to have been derived from those braces with re-curved upper ends that had been used during the reign of Henry VII. It may be seen on the façade of the house built for Henry VIII in the Tower of London in 1528, where the braces form 'S' curves that had little strength because they were cut across the grain of the timber. This fashion spread despite its costliness and decadence, and in streets of wealthy merchants' houses, such as may be studied in Coggeshall's Church Street, examples of completely 'serpentine' wind bracing can be seen, some of it dating to the 1590s. In closely-spaced urban timber building, however, such as the residential range of Staple Inn, High Holborn, wind pressures had become less of a factor to be taken into account, and such façades had no bracing, although some was retained in the cross walls to counteract the imbalance of the asymmetrical jettying. By the end of the century (as shown by Rooks Hall, Fig. 205), the wind bracing which was still necessary in open country had been reduced to the functional minimum and was placed in the four corners of houses at first-storey level. In this phase it has become a primary constituent of the frame and was chase-tenoned into place when the carcase was reared; the studs had become of secondary importance, being cut where they met the braces, and nailed to them.

Appendix Five

Posts having carved capital and base treatments

Fig. 324

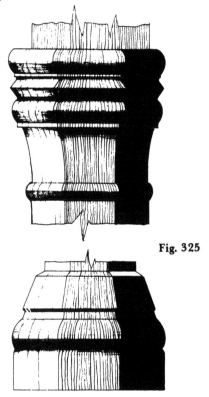

Fig. 325

between *c.* 1080 (as at Worcester), and *c.* 1150. The later of the two dates is suggested by Mr. Rigold (*Medieval Archaeology*, forthcoming).

Fig. 324. Capital treatment from the freestanding posts of the prebendal hall at Faulkners, Good Easter. This may be described as a 'cubical' capital, which Clapham called the 'normal type of Anglo-Norman capital . . . commonly called the cushion-capital'. Its most distinctive feature is the carinated fillet which had a widespread and persistent usage during the vogue for the Romanesque style. In its accentuated form, with the soffit plane larger than the upper and narrow plane, it appears in the late 11th-century crypt of Worcester Cathedral; and in the symmetrical form with equal facets it occurs at Durham, Selby, Peterborough, and Rochester during the 11th century, thereafter persisting in both forms in the choirboys' vestry at Winchester Cathedral until *c.* 1180. In this form, without attached roll mouldings, it should be of early date, i.e.,

Fig. 325. An 'arcade' post from Crepping Hall, Wakes Colne, Essex, probably dating from between *c.* 1130 and *c.* 1180. This capital illustrates perfectly what Mr. Rigold called 'the almost carpenterly delicacy of East Anglia', being precisely that. It is probable that few freestones ever admitted of the fine and accurately-cut details of this carinated fillet, which may be a carpenter's version. Such fillets had a protracted usage during the Romanesque years throughout England, and the date range extends from the late 11th-century crypt of Worcester Cathedral until the final decade of the 13th century at Wells Cathedral.

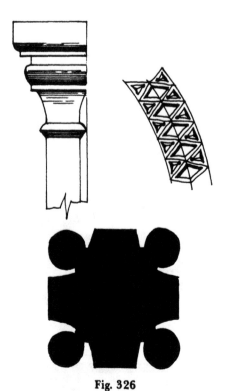

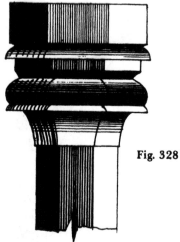

Fig. 327. From a queen post of Gatehouse Farmhouse, Gransmore Green in Essex. The water-holding base is indicative of a date during the 13th century, and the section of the capital suggests the central decades of that century, conforming as it does to the half-roll with frontal fillet motif to which Forrester ascribed a period of maximum popularity between *c.* 1229 and 1300.

Fig. 328

Fig. 326

Fig. 326. From Great Bricett Priory in Suffolk, probably dating from between *c.* 1210 and *c.* 1250, since associated with dog-tooth ornament.

Fig. 328. The capital of the first-floor samson post from King John's Hunting Lodge at Romsey, Hampshire. It was not, of course, a hunting lodge; neither does it resemble such a building, and this samson post was not, as is sometimes suggested, a former crown post from its roof. The datable motif is again the half-roll with frontal fillet.

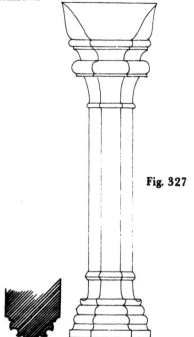

Fig. 327

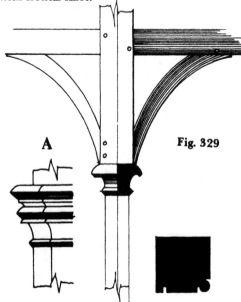

A

Fig. 329

Fig. 329. A capital, half in elevation and half in section, from a quasi-king post in the roof of the service wing at Tiptoft's Hall, Wimbish, Essex. At 'A' in this figure is the semi-octagonal profile of one of the capitals from the four jamb posts provided for the original three service doors; this shows more clearly the important profile, which is cut three times on each example (the pointed roll). This, as Forrester stated, 'affected moulding work but little, the keeled roll being preferred. Existing examples are rather rare' (Forrester, 1972, 12). The period of use was *c.* 1200 until 1220, but two decades. In view of this degree of rarity and the absolutely perfect execution of these oft-repeated profiles, a date within the period of their use is proposed for the service wing.

Figs. 331 and 332 (p. 300). The crown posts from Kennington's at Aveley in Essex. These can only be dated by the use of accurately-cut scroll mouldings to between *c.* 1260 and 1340.

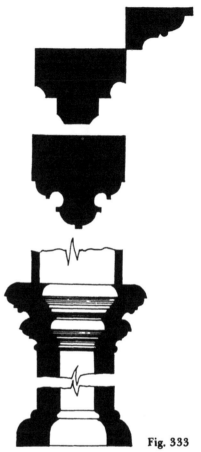

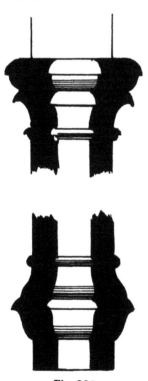

Fig. 330

Fig. 333

Fig. 330. The crown post from Place House at Ware, in Hertfordshire. This has been ascribed to *c.* 1295 by Forrester, and its definitive constituent is the tongue-shaped adjunct to the roll, which is datable to between *c.* 1250 and *c.* 1320. The strongest evidence as to the actual date is the tongued-and-grooved scarf joint of the top plate, which points with tolerable certainty to the reign of Edward I (1272–1307).

Fig. 333. Capital from arcade post of Stanton's Hall at Black Notley, Essex. This again is datable by the emphasis upon scroll mouldings; three-quarter-circle hollows are combined with the roll with three fillets. The scrolls date from between *c.* 1260 and 1340; rolls with three fillets from between *c.* 1270 and 1330, and three-quarter-circle hollows from the same period. As previously mentioned there is documentary evidence for a dispute in 1306, and a date *c.* 1300 is proposed for the building. The base mouldings of these main posts closely resemble those of the south arcade piers of Navestock church (C. A. Hewett, and J. R. Smith, 1972, 830), which was dated to *c.* 1250 by the Royal Commission in 1921, thus providing evidence for a date earlier than *c.* 1300, as do the joist tenons placed nearer the soffits than the centres of those timbers.

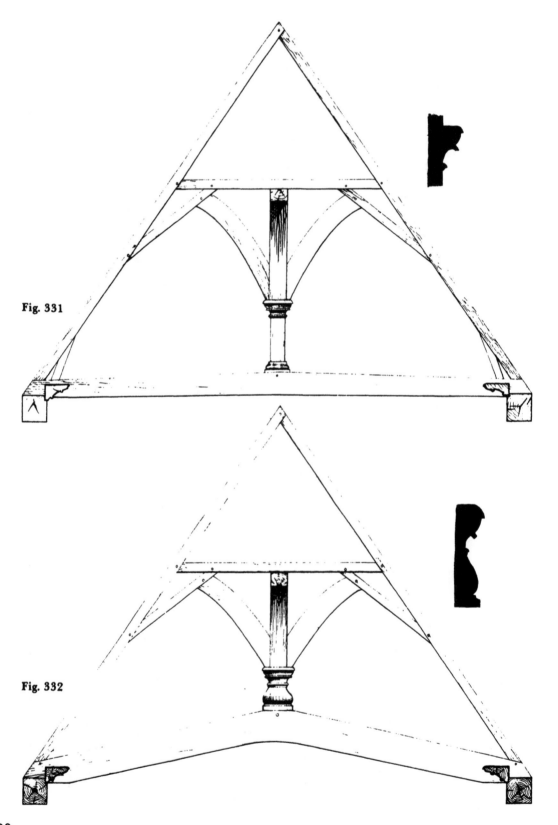

Fig. 331

Fig. 332

300

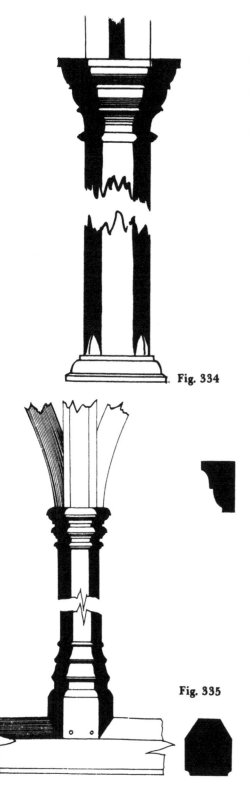

Fig. 334

Fig. 335

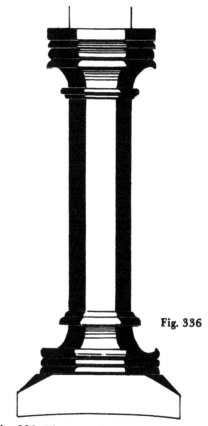

Fig. 336

Fig. 334. The crown post from Southchurch Hall, Essex. Tongue-shaped roll adjuncts place this between *c.* 1250 and 1320; a date close to *c.* 1300 is suggested.

Fig. 335. The crown post from the nave of Pattiswick church in Essex. The use of three scroll mouldings places this in the date range of that profile (*c.* 1260 to 1340). As a feature the bell-spread base is generally ascribed to the Perpendicular style, but in timber it appears earlier, the earliest examples being placed above the base of the post (as at Place House, Ware).

Fig. 336. The central crown post from Baythorne Hall in Essex. The datable constituent here should be the water-holding base, which, according to Forrester, 'went out of favour about the middle of the thirteenth century', and persisted 'in isolated cases to the end of the thirteenth century'. The only technological evidence in this building is the use of soffit tenons for the first floors, resembling the choir stall floors of Winchester which were completed by 1309. The early Decorated period is suggested for this house, but within that period the date must be conjectural.

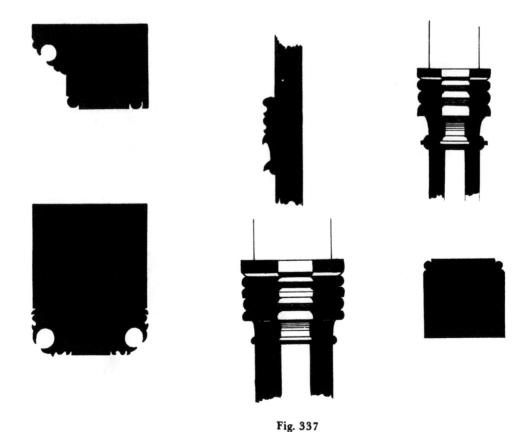

Fig. 337

Fig. 337. A selection of capitals and other profiles from the 'stables' at Church Farm, Fressingfield, Suffolk. The most expensive and impressive to view among these are the three-quarter-circle hollows, of large diameters, and dating between *c*. 1270 and 1330. Within this range the building must be placed, but no other evidence is known to date.

Fig. 338. An arcade post from the second phase of construction of Great Chesterford Manor House, which is the aisled timber hall. The sections of these are quatrefoils, each treated as rolls with three fillets, and date from between *c*. 1270 and 1330 (of which the latter is usually suggested). The capitals have two ogee scrolls combined with two of the hollowed rings noted in carpentry, but not specifically mentioned by Forrester. These embody a tongue-shaped roll-adjunct in both cases, and occur on the capitals of Hereford Cathedral in *c*. 1260 (H. Forrester, 1972, 42).

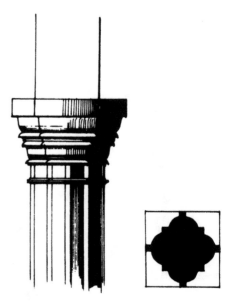

Fig. 338

302

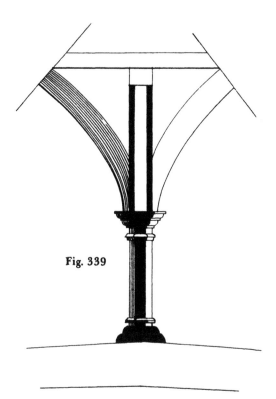

Fig. 339

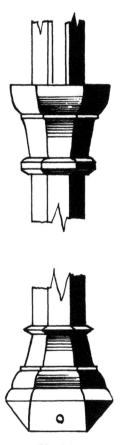

Fig. 341

Figs. 339 and 340. Capital profiles from Lampett's Farmhouse and St. Clere's Hall, both in Essex. In both cases octagonal sections were used, and the mouldings bear a slight resemblance to the 'wave' profile, but must not be confused with it. They are known to date from *c.* 1350, which is the final date for waves; this form, however, seems more common in timber than in stone.

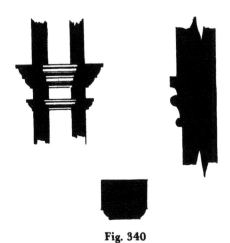

Fig. 340

Fig. 341. The crown post from Bridge House at Fyfield in Essex. Its bell-spread base indicates the Perpendicular style, whilst the sharp knife-edged ring about the base is a feature frequently found in timber examples, deriving from an early use at Ottery St. Mary, between 1336 and 1342. It is proposed that this specimen dates from the 14th century.

Fig. 342. The crown post from the northern cross wing of the *Woolpack* inn at Coggeshall in Essex. This exists in association with a soffit-tenoned first floor, for which reason it belongs typologically to the 14th century— possibly to the beginning of that century. The only decorative treatment is the stop chamfering, which does not assist with dating. The same type was used until late in the 16th century, when its components were reduced to thin, plank-like sections. As an anti-rack measure it was more efficient than other types, because

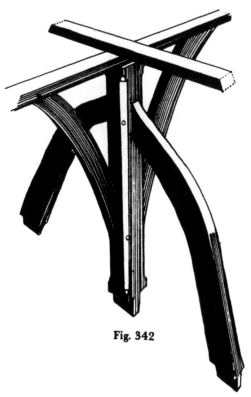

Fig. 342

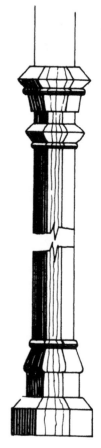

Fig. 343. The central crown post from Durham House, Great Bardfield, Essex. A tentative date ascription for this building rests on the evidence of the head-dress carved on the woman's portrait on a hammerbeam end, and nothing conflicts· with that ascription of *c.* 1370. It could, of course, be earlier in date. The crown post is of interest because it is not datable by any known criteria, although it is treated in an architectural and decorative manner.

the collar purlin braces reached low down the post, offering more objection to inclination from the vertical.

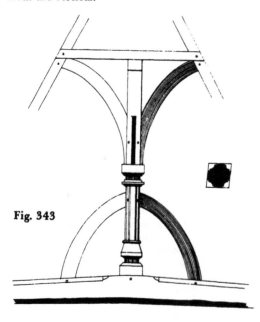

Fig. 343

Fig. 344

Fig. 344. A crown post from the nave roof of the parish church of High Ongar in Essex. These are tall and octagonal, and illustrate a peculiar profile of both abacus and astragal. According to R.C.H.M. (1926, 131) the nave dates from *c.* 1150, and was last fitted with two new windows early in the 14th century. Forrester obšerved that the base profile was a frequent element in 15th-century timber profiles, and suggested the first half of that century.

304

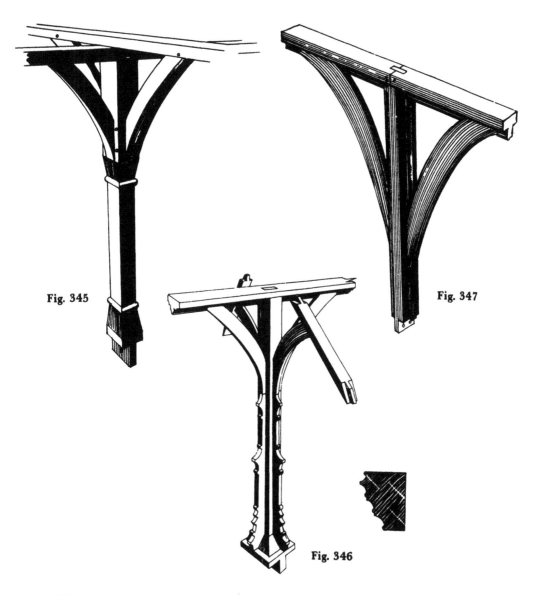

Fig. 345

Fig. 346

Fig. 347

Fig. 345. The crown post from the house named Tudor Cottage in Stebbing, Essex. This, again, is an undatable artefact, the appearance of which is suggestive of the Romanesque. It can be dated by reference to the frequent scarfing of the top plates of the building, which are edge-halved with bridled abutments, not known before c. 1375-80, when they were used at Trig Lane, London. The date must, therefore, lie between c. 1375, at the earliest, and c. 1595, the date of the last known crown post in the county: at Makrons, Ingatestone, Essex (Hewett, 1969, 145).

Fig. 346. An example from the north aisle of the parish church at Stock Harvard, Essex. This is of the cross-quadrate section, and cut to profiles peculiar to the Perpendicular style between c. 1350 and 1450.

Fig. 347. Crown post from Barnard's Inn, London. It is allegedly of the later 14th century, an ascription with which no structural details are in conflict. As an example from London it is of great interest and rarity.

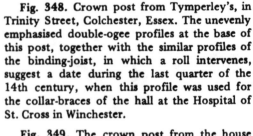

Fig. 348. Crown post from Tymperley's, in Trinity Street, Colchester, Essex. The unevenly emphasised double-ogee profiles at the base of this post, together with the similar profiles of the binding-joist, in which a roll intervenes, suggest a date during the last quarter of the 14th century, when this profile was used for the collar-braces of the hall at the Hospital of St. Cross in Winchester.

Fig. 349. The crown post from the house known as Monk's Barn at Newport in Essex. It was never, needless to say, a barn, nor anything resembling one. The uncommon quality of the carving is remarkable for two features; the 'inverted-wave' profile under the abacus, to use Forrester's term for it (H. Forrester, 1972, 29), and the clearly-cut hollow moulding round the base of the bell foot. The former is paralleled in a transept screen at St. Cross Hospital in Winchester, while the latter was used for the capital of the wooden central pier of the chapter house roof at Salisbury Cathedral of *c.* 1275. Both Mr. Forrester and Dr. Harvey recognise the emergence of the inverted-wave profile during the early 14th century (J. H. Harvey, 1978, 36). This profile appeared in the south transept of Gloucester Cathedral, *c.* 1331, when William Ramsey was possibly a consultant.

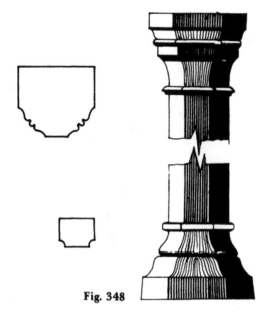

Fig. 348

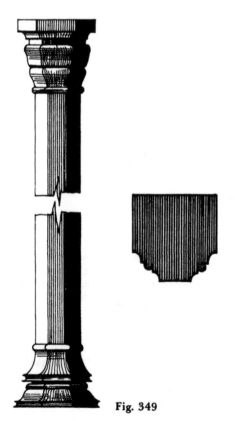

Fig. 349

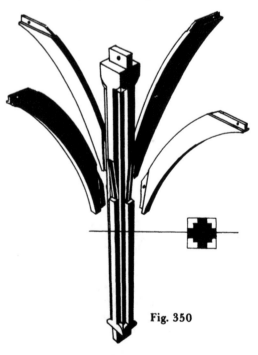

Fig. 350

Fig. 350. A crown post from the houses associated with Dorothy Sayers in Newland

Street, Witham, Essex. This is of the cross-quadrate section which had very early origins, being adapted to profiled 'arms' during the Perpendicular period, than which it had a longer period of use, or advocation. The floor joists of this building indicate a date between *c.* 1400 and *c.* 1500.

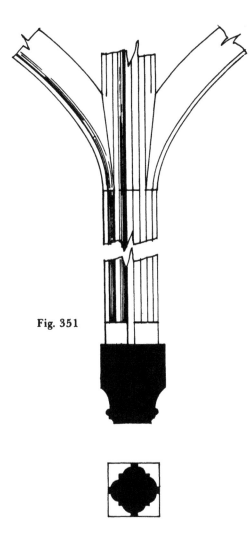

Fig. 351

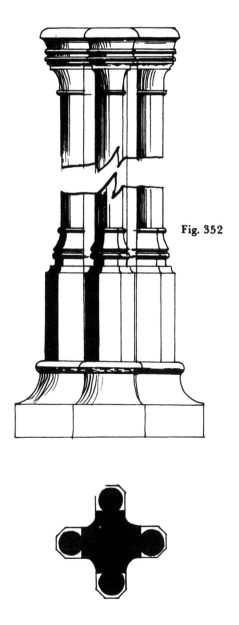

Fig. 352

Fig. 351. Crown post from the chancel roof of the parish church at Tolleshunt D'Arcy in Essex. This church looks Perpendicular in style; the cross section of the crown post, however, is the same as the arcade posts of Little Chesterford Manor House, and that of the tie beam is a beaded casement. A date during the 14th century is proposed.

Fig. 352. Post from the timber north arcade of the parish church of Shenfield, Essex. This is not dated, but the closest parallel among Forrester's examples is from Ewelme, Oxon., *c.* 1450.

Figs. 353 and 354. Two crown posts, one from the hall of the *Woolpack* inn at Cogges-hall, and the other from the hall of Jacobe's Hall at Brightlingsea, both in Essex. The first is of similar date and style to the second, which was ascribed by Forrester to *c.* 1460–70 with

307

some apparent precision; the structural features of Jacobe's certainly agree with such a date. The resemblance of the capital at the *Woolpack* to a crown or coronet is possibly misleading, since it may result from the application of the 'crenellation' ornament to an octagon. The ornament seems to appear first in the vicars' houses of the Vicars' Close at Wells, some of which were completed by 1348.

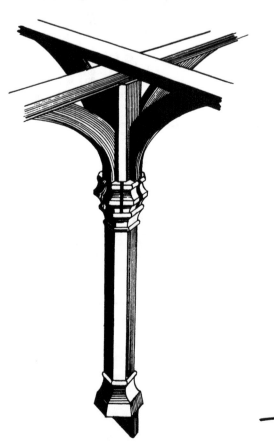

Fig. 353

Fig. 354

Appendix Six

String- or band-type mouldings

Fig. 355

Fig. 357

Fig. 355. A major rib profile from the timber 'vault' of the presbytery at St. Albans, the 'earliest surviving example of its kind on a large scale' (R.C.H.M., 1952), built between 1235 and *c.* 1290 (J. H. Harvey, 1976, 155). This profile had to be taken from the full-size drawings on site dating from the last restoration, but is assumed to be correct. The profiles have pointed rolls, which according to H. Forrester were popular between 1200 and 1220.

Fig. 357. A rib profile from the timber 'vault' of the south choir aisle at Rochester Cathedral, ascribed to *c.* 1322 by St. John Hope (St. J. Hope, 1900, 82). This emphasises the roll with three fillets, which dates from between *c.* 1270 and 1330; the plain roll set within a large hollow is of interest, and reappears often in the timber mouldings to follow.

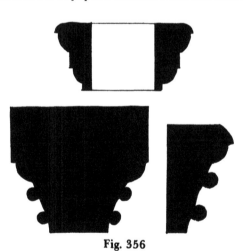

Fig. 356

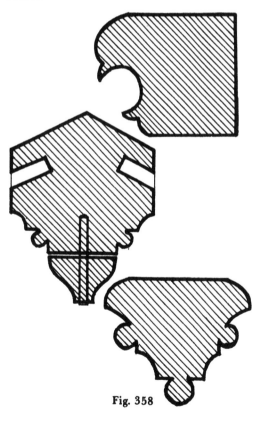

Fig. 356. Profiles from Place House at Ware, of *c.* 1295. The important profiles here are the scrolls, which are datable to between *c.* 1260 and 1340; the use of small roll mouldings seems to have been persistent among carpenters from early times, but there are also parallels in stone, as in a tower arch of *c.* 1300 at St. Mary Redcliffe.

Fig. 358

Fig. 358. Three timber sections from the nave roof of the parish church at Sheering, Essex. The wall plate (uppermost) features the three-quarter-circle hollow and tongue-shaped adjuncts, which together range from between *c.* 1250 and 1330. The ridge piece (centre) has the roll with frontal fillet, with a date range of *c.* 1220 to 1260; but in timber mouldings an ogee curve just as frequently connects fillet and roll. The wall corbels again feature rolls set in major hollows. The structural tenons on the rafters of this roof are barefaced soffits, and an early 14th-century date is proposed.

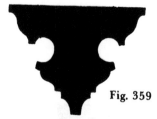

Fig. 359

Fig. 359. A section through a 'built' tie beam in the nave roof of Goldhanger parish church in Essex. This shows both wave mouldings and three-quarter-circle hollows, and probably dates from between *c.* 1310 and *c.* 1350.

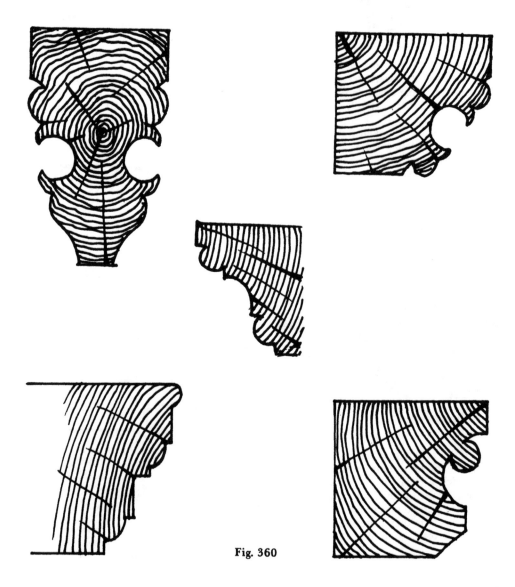

Fig. 360

Fig. 360. A selection of profiles from timbers in the Guard Room roof at Lambeth Palace. Among these, which include the three-quarter hollows and the tongue-shaped adjuncts to the rolls, appears the roll within a hollow again. The date range should be determined by the first two features and covers the period between *c.* 1250 and *c.* 1330.

Fig. 361

Fig. 361. The section of the internal wall plates at Aythorpe Roding parish church in Essex, the dominant profile being the roll with three fillets. The date should lie between *c.* 1270 and *c.* 1330.

Fig. 362

Fig. 362. The section of an internal eaves-cornice timber, from Brett's Hall at Aveley in Essex. This timber is not necessarily *in situ*, and strictly comparable stone examples of rolls with three fillets are not easy to trace; it probably dates from the period between *c.* 1270 and *c.* 1330.

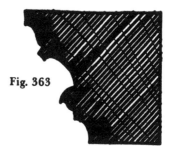

Fig. 363

Fig. 363. A wall-plate section for the parish church at Messing, Essex. This is datable by an armorial achievement to between 1344 and 1362. It incorporates an ogee scroll of 14th-century style, set into a major hollow, which presages the casement.

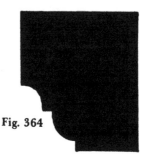

Fig. 364

Fig. 364. A profile from the *Old Sun* inn at Saffron Walden. This was dated in its whole context by Forrester to between *c.* 1350 and 1360. Other than that it indicates a Decorated preoccupation with circle segments, it is not a datable profile out of context.

Fig. 365

Fig. 365. The principal section used at Barnard's Inn, which is of the late 14th century. A form frequently to be encountered in timber mouldings, in effect a casement with a roll within and hollows above and beneath, it seems to persist with variations from *c.* 1370 until *c.* 1550.

311

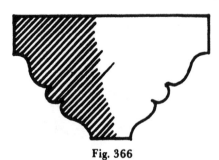

Fig. 366

Fig. 366. The soffit section of the collar arches in the Brothers' Hall of the Hospital of St. Cross at Winchester. This is dated to 1372 by the Hospital's records.

Fig. 367

Fig. 367. From the internal wall plates of the parish church of High Roding, in Essex. This is a pronounced example of a three-quarter-circle hollow with a roll within, now almost reduced to a bead; the quarter-hollows are retained above it and beneath. Mr. Forrester observed: 'this seems to be characterised by segments of the circle on the chamfer-plane, a feature most noticeable in stone in earlier 14th-century work. Allowing for the intrusion of the small bowtell, the wall-plate may be of late 14th-century date'. The scarf joints used are edge-halved with bridle abutments, and a date a little before *c.* 1400 is proposed.

Fig. 368

Fig. 368. The section of the internal wall plates of the parish church at Leaden Roding in Essex. Notable features are the wave moulding and the beaked half-roll, and of the wave form Forrester observed that 'the stage [shows] some exaggeration of the concaves and a consequent diminution of the entasis', suggesting a late 14th-century date.

Fig. 369

Fig. 369. Mantle beam profiles from Fryerning Hall in Essex, ascribed to the period between *c.* 1425 and *c.* 1475. In this case the tools used were very good and well sharpened, and the resulting profiles are symmetrical. Inferior work is more usual during the 15th century.

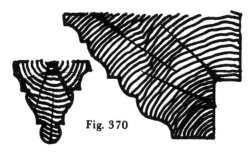

Fig. 370

Fig. 370. Section of a durn and a muntin, from the doors to Charterhouse Square.

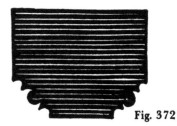

Fig. 372

Fig. 372. The cross section of the binding-joist in the first floor of the Great Standing at Chingford, Essex. This was completed as a building in 1543, and the choice of sections and their production would not have been much earlier, since they were King's work.

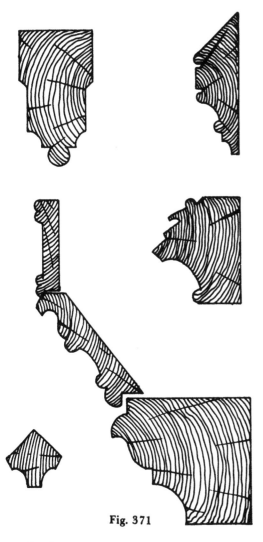

Fig. 371

Fig. 371. Moulded cross sections from the Old Hall at Lincoln's Inn, dated 1493.

Fig. 373

Fig. 373. The section of the mid rail of the north transept screen, Hospital of St. Cross, Winchester, together with that of one mullion. The rail has two 'inverted waves', a distinctly Perpendicular feature.

Fig. 374

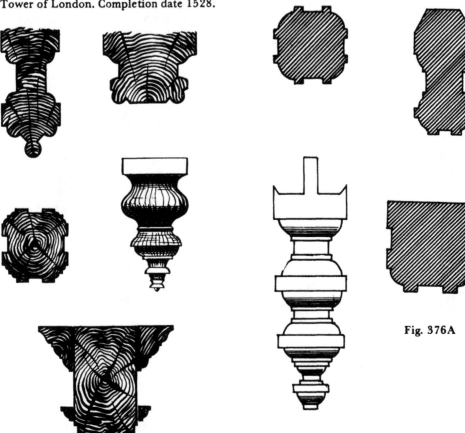

Fig. 374. The cross section of the attic-window stiles which support Tudor, pointed, four-centred heads at the Queen's House in the Tower of London. Completion date 1528.

Fig. 376A

Fig. 375

Fig. 375. Cross sections from the roof of the Great Hall at the Middle Temple (the former was begun, and therefore designed by, 1564). The sunken channel is most important, as are the 'built' hammerbeams and the turned pendants. The preference for quarter rounds within returns suggests the preference to come for ovolo sections.

Fig. 376. At 'A' can be seen some cross sections from the roof of the Great Hall of Staple Inn, High Holborn, together with one of its pendants, and at 'B' sections from the residential range along the Holborn frontage at Staple Inn. The obtusely-angled section is that of the corner mullion of the oriel window, which incorporates both sunken channels and ovolos, the latter in the form that was to predominate during the early 17th century.

314

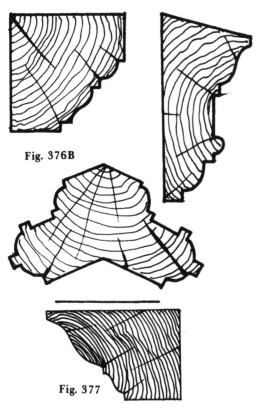

Fig. 376B

Fig. 377

Fig. 377. One important moulded section
from Bishop Juxon's Great Hall, now the
library, at Lambeth Palace. This is a major and
a minor ogee set on a chamfer plane of 45degs.
Dated to 1663, this section was to become
invariable in the 18th century for internal
cornices of timber.

Fig. 378

Fig. 378. A section from the dated porch
of the parish church at Sutton in Essex, of
1633. This could date from the preceding
century, incorporating as it does profiles
frequently combined during the 15th century.
It serves to illustrate the subtle differences
between real and eclectic profile combinations.

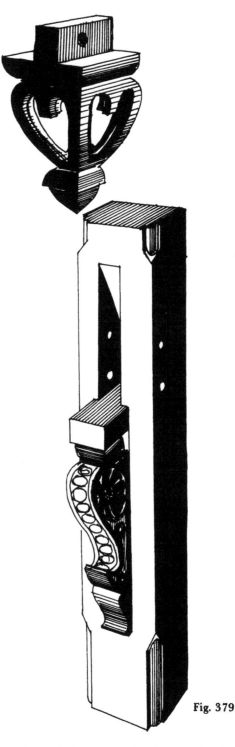

Fig. 379

Fig. 379. Decorative details from the dated
roof of Dore Abbey (1632).

315

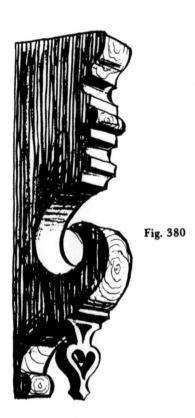

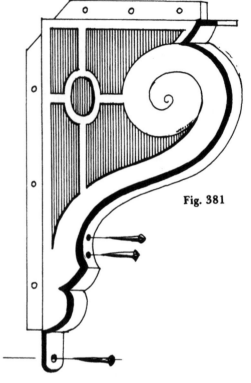

Fig. 380

Fig. 381

Fig. 380. An elaborately-profiled hanging knee from a house in Halstead, Essex. Dating from the middle years of the 17th century, this incorporates, at necessarily great expense of time and labour, the pierced pendant motif which is present in the Dore Abbey roof details. The shallow and repetitive profile at upper right is from the tie beam of the same house.

Fig. 381. A hanging knee from Chipping Ongar, Essex, aptly described by Sir N. Pevsner as a 'scrolly bracket'. This illustrates the realisation that such wide knees, when cut (as distinct from grown bends or conjunctions), were weak; and it was bound with iron to correct this weakness. This house is dated 1642.

Fig. 382. An apex and a valley conjunction from the verge boards of the Queen's House, Tower of London, dated 1528. These are of great interest since they illustrate the taste—still Gothic—of the capital in that year, when Tudor arched timber windows were fitted into this same building.

Fig. 382

The development of storey posts

During the early stages this was concurrent with the development of integral first floors and jettying, and the posts of both barns and houses were little more than mere posts, with several joint cavities cut into their various surfaces. In ancient systems, such as that surviving at Paul's Hall, only the tops had tenons, the feet being plane surfaces; this was also true of the Grange Barn of Coggeshall Abbey. Such unspecialised main posts could, and often did, receive elaborate decorative treatment as piers with capitals and bases, without any thickening or diminution of their volume other than that indicated by their ornament. This is illustrated by the posts of the Bishops' Palaces at Farnham and Hereford, Crepping Hall and Faulkner's Hall.

By the middle of the 13th century, when the Cressing wheat barn was built, upstands were provided at the posts' tops, in order that the three main structural members of framed buildings might be jointed as integrated assemblies, instead of as one member to another in an additive succession. This was the first step in the development of specialised posts for timber buildings. By the end of the century (as is illustrated by Place House at Ware) the upstand had become the jowl, and threefold jointing—in which each of the three timbers was physically attached to the other two—perfected: no longer would the failure of any one of them mean collapse. With the increased production of storeyed timber buildings during the 14th century, and an ever-increasing awareness, due to failures, of the fact that storey posts with little of their substance left at side-girt level would often break, the specialised storey post developed. This took the form of a post with a jowled top and another jowl at first-floor level, the lower being intended to provide a ledge or bearing for the support of the binding-joist. This, however, was still a weak point in such posts, because mortising was necessary in two opposite edges, and also in their internal faces at the same base level. The service wing of St. Clere's Hall (Fig. 156) contains a storey post designed with two upper jowls facing in opposite directions at two different levels, in addition to a third which was left as an arch brace impost. This example dates from c. 1350. From that time and until the early 16th century, the development of floor framing continued along with the appropriate forms of mortises and tenons, and storey posts elaborately designed for specific purposes were normal. During these years the weakness of the inevitable triple mortise at floor level was counteracted by greatly increasing the widths of the posts, which frequently made them wider than their tie beams. This fact seems to have brought about the treatment of the posts above the first floors, as of two square orders; in such cases the inner and narrower order fitted the width of the tie beam. An example, based upon the posts of Childe's Chantry, is shown inset in Fig. 183. By the later years of Elizabeth I's reign storey posts had begun to revert to mere posts, and such houses as Doe's at Toothill had but a single pair of jowled posts in their framing, with no additional thickness at their weak floor-level points.

The development of mortises

The mortise, or squared hole, is so basic and essential to carpentry that it might be thought that it had always had the same appearance as that with which we are familiar, if we are accustomed to viewing mortises. This is not the case, however, and a long course of development can clearly be traced, ending in highly complex multiple mortises such as may be found on the lock rails of Georgian doors. It has been established that mortising was practised for framing houses in Neolithic Europe, at Köln-Lindenthal and Aichbühl (Dr. N. Davey, 1961, 36), and thereafter through the Iron Age until present times. It is not necessary to prove the origin of the mortise before suggesting that its earliest form must have been a circular hole made by some form of drill, and then squared with some form of chisel, both bone and flint being possible materials for such tools. Moreover, the removal of most of the waste wood from mortises by means of drilling with shell augers continued to be standard practice throughout medieval times. It is demonstrable that most mortises in ancient timber structures, such as the barn at Paul's Hall or the Grange Barn at Coggeshall, have a low aspect ratio when viewed on plan; their lengths are little greater than their widths, a fact which relates them to their suggested origin. From the 11th century onward this joint has been improved, by multiplying the cavities and spacing them evenly across the timber face, by avoiding through mortising, and by providing various forms of sunken butment cheeks to accommodate either spurred shoulders or diminished haunches. This process has only been outlined but it is implicit in various buildings that are illustrated, one very good 16th-century example being the Great Standing at Chingford (Fig. 199), where a pair of single tenons was used to resist extension and their three shoulders were spurred to resist shearing.

Some technological changes illustrated by the examples

c. 1026 The barn at Paul's Hall, Belchamp St. Paul's, lime cement-based posts, apparently reared against earthfast shores.

c. 1140 The Grange Barn, Coggeshall Abbey, stone stylobates and framing fully independent of its site. Refined profile notched-laps.

c. 1175 Cressing barley barn, groundsills secured to the aisle walls.

— 1209 Wells Cathedral, high roof, open notched-laps.

+ 1213 Wells Cathedral, high roofs, secret notched-laps.

c. 1230 Salisbury, spire scaffold, splayed scarf with bridled abutment.

c. 1250 Joists' end tenons set beneath the centre lines of joists.

c. 1250 Cressing wheat barn, the upstand, the splayed, tabled and keyed scarf, square under-squinted butts.

c. 1250 Salisbury, Old Deanery, the tongued-and-grooved splayed scarf.

c. 1295 Place House, Ware, splayed, tabled, keyed and sallied scarf with tongued-and-grooved tablings.

c. 1299 Fyfield Hall, existing 'new' roof, paired principal-rafters on tie beam's outer edges, scissor-braced.

c. 1307 Winchester Cathedral, choir stalls, barefaced soffit tenons.

c. 1340 Vicars' houses, Wells Cathedral, single tenons with soffit spurs, scribed. The crenellated ornament.

c. 1350 St. Clere's Hall, scarf with stop splay and two bridled abutments.

c. 1375 Trig Lane waterfront, edged-halved scarf with bridled abutments.

c. 1398 Limpsfield Court House, possible use of barefaced soffit tenons in alterations for the abbott of Battle.

c. 1404 Beaufort Tower, St. Cross Hospital, Winchester, single tenons with housed soffit shoulders.

c. 1420 The Chequer, Wells Cathedral, roof with inverted wind braces above side purlins, derived from tower arches. Soffit spurs to side purlins' single tenons.

c. 1433 Bishop Bubwith's library, Wells Cathedral, single tenons mounted on spurred shoulders.

c. 1434 Lambeth Palace, Chichele's Tower, scribed joists with central single tenons.

c. 1493 Lambeth Palace, Morton's Tower, scribed common joists.

c. 1500 Paycocke's House, central single tenons with housed soffit shoulders still in use.

c. 1510 King's College Chapel, soffit tenons with diminished haunches.

c. 1528 The Queen's House, Tower of London, deep-sectioned joists, combined with humped tie beams in a flat attic floor. Façade gables.

c. 1575 Rooks Hall, Cressing, face-halved and bladed scarf.

c. 1595 Clintergate Farmhouse, counter-bladed scarf.

c. 1650 Barn at Rickling Hall, edge-halved and bridled scarf used.

c. 1677 The *George* inn, Southwark, soffit tenons with diminished haunches used on softwood framing.

Carpenters' mouldings

Little has been published concerning these, for the very good reason that they have not been collected, classified, and dated by comparison with masons' mouldings, to which they are directly related. This fact reflects the curiously limited attitude of the earlier 20th-century authorities to matters of architectural history, a subject on which a great deal was then published. The vast amount of research that would be entailed in a definitive work on timber mouldings during the Gothic period must await the attention of some one who is a professional archivist, an architect and a craftsman; but the selection of examples described does suggest a basis for such studies.

It is apparent that no moulding profile among those illustrated is peculiarly suited to execution in timber, rather than in stone; and it is equally apparent that good timber facilitates the cutting of sharp-edged projections that would seldom prove either possible or durable in building stones. The composition of stone and timber is entirely different, stone being granular and timber fibrous, and this could have led to distinctly different series of mouldings; but in fact it produced subtly varied versions of the same series. Mouldings, therefore, owe their origins to the prevailing styles of English architecture, and their dating by means of cross references between masonry and carpentry is valid, since both expounded the same single history of national decorative style. There are no grounds for the still prevalent assumption that masons came first and received information before carpenters, who have generally been held as belated emulators of their fellow tradesmen (a myth relating to the traditional view that man must have built in stone before timber). Both of these views become ridiculous when considered seriously. Had two of the greatest of carpentered works—the Ely Lantern and the roof of Westminster Hall—been designed at times when innovations were due in the history of ornamental profiles, then such innovations would have found their first expression in timber. That this did not happen in either case indicates the overall expression of Medieval style in all craft media concurrently, without special emphasis on any one.

The methods used by craftsmen to produce profiles on timbers can, for the most part, be deduced from the nature of the tool marks left; 14th-century roll mouldings were frequently scratch-stocked, a method leaving evidence of 'chatter' in the form of minute steps placed squarely across the moulding. Such mouldings inevitably follow the grain because the tool has no sole, which is why the earlier mouldings have sinuous courses along the timbers.

With the availability of moulding planes it is immediately apparent that quite elaborate profiles were cut straight, right through any patches of short grain. The complex profiles carved on the capitals and feet of both crown posts and arcade posts in halls were frequently paired, across the grain, with gouges and chisels; but in the best work no discernible tool marks remained. Straight chamfers and basic chamfer planes were produced with draw knives, often after the building was erected; and during the period of the three-quarter-circle hollow this plane must have been the starting point. Forrester maintained that the period of most frequent usage was from c. 1270 until c. 1330, but it has not yet proved possible to deduce the tools with which these were cut. In the decadent examples (such as those in the Fressingfield building) large circular hollows, perhaps three inches in diameter, were cut, but leaving only a one-inch opening along the edge, thus serving only to produce an emphatically dark line without displaying any of the hollow contour. In the Winchester choir stalls such hollows were worked round inside curves, when no form of planing would have been possible; but when the coopers' groze—a form of plane producing the indent around the inside of a barrel's staves, into which its ends are fitted—was first used is unknown, and coopering was certainly practised as early as the Saxons, so some specialised hand tool remains a possibility for this purpose.

It appears from the carpenters' profiles illustrated that the casement profile, which characterised the Perpendicular style, was derived from the three-quarter-circle hollow of the preceding Decorated period. The section in Fig. 367 represents the rare occurrence of a bed or bowtell inside such a deep hollow, and if the hollow is made shallower this combination results in the beaded casement. The cited example is late 14th-century, but it may be illustrative of the process.

Contrary to first impressions roll mouldings in close series were frequently used, at least from c. 1215 at Little Coggeshall until the last years of the Tudor style. These can readily be produced by hollow planes, which have a sole section that is concave; and if such planes are bull-nosed the rolls can be planed up to the point where the 'stop' is to be carved. It is probable in the light of the Lambeth Palace examples (Figs. 297 and 301) that speed and cost of execution influenced design, because no chamfer stops were needed and the rolls were planed straight through and off the work at both ends. A parallel process can be discerned in the evolution of the 'creased' or 'shadowed' door leaves of Elizabethan times. Early examples have the cross-sectional contours of the preceding linenfold motif which was produced with a number of hollow and round planes; if the stops were not cut the full-length contours produced the creased door.

There are no grounds for duplicating the late H. Forrester's work on mouldings, and his book is an essential aid to a study of the subject (H. Forrester, 1972). Forrester's contribution to the study of mouldings was very important, and he attempted to define the periods of maximum popularity, or usage, of their definitive elements; but it must be remembered that he gave examples of each which are dated both earlier and later than these suggested periods of popularity. In the case of the truly great houses such as Tiptoft's, Chesterford Manor House or Stanton's Hall, to name just three, it must also be born in mind that the use of a moulding element before the period of its maximum popularity is as probable as its use during that period. Houses with royal associations such as the Bury at Clavering can be expected to employ incipient moulding elements, and this applies equally to those of prelates, aristocrats and knights of the calibre and importance of Sir John de Wantone, who was associated with Tiptoft's.

The succession of roof designs

There has been more speculation in the past on this matter than upon others that have already been discussed, but it has not previously been possible to make any appropriate allowances for events before the Conquest, for which some hazy indications have now been established. Faced with the magnitude of All Saints' church at Brixworth, Northamptonshire, with a nave spanning almost forty feet, most serious scholars have persisted in the belief that there could have been no tradition of large-scale building in England during the centuries between the Roman and the Norman influences. Brixworth unfortunately seems to provide less evidence than smaller surviving examples, but its size alone compels one to acknowledge the considerable skill involved in such timber roofs; it may have been built for or by King Ethelbald of Mercia, c. 750 (The Rev. N. Chubb, 1977, 3). The techniques of the masons involved are definitely unskilled; the turning, for example, of the nave arch now forming the south door is unscientific by comparison with later, voussoired arches. However, it is suggested that craft skills in the mid-eighth century had derived from the uses of timber and metals and other attractive materials, a bias that is necessary to account for the development of fine-cut ashlar and well-burnt limes as late as the 11th century. The masons' craft was, in fact, among the more recent developments, as has been stated: 'The earlier Norman masons were not yet fully competent to design the large structures which they were attempting'. (J. H. Harvey, 1971, 21).

A remarkable degree of imbalance is demonstrated by comparing the levels of skill required for building the Sutton Hoo ship, or the production of its grave goods and jewellery, and any work of early Romanesque masonry, including early Norman masonry. In the absence of any sufficient evidence it has to be assumed from the Brixworth nave that eighth-century carpenters could build roof spans of over thirty-five feet, with framing that evidently did not overturn the inadequate stone walls—because the two major crafts, masonry and carpentry, although inextricably inter-dependent during the Gothic periods of building, did not have a common course of development and did not evolve simultaneously.

The church of St. Martin at Canterbury, the spire at Sompting, and the chapel at Harlowbury are surviving buildings which suggest, during the Late Saxon period and for some time after the Conquest, the construction of timber roofs over carcase walls that were finished at the pitch of the rafters. In the case of Harlowbury chapel those rafters were actually embedded in the rubble, which would also appear to be true of the other two (subject only to examinations when repairs make them practicable). In two cases, St. Martin's church and Harlowbury, the collar was apparently used in extension and such roofs were tied at collar level, a supposition advanced by Davey (Dr. N. Davey, 1961, 45), who observed that the 'collarbeam was possibly the first structural member designed as a tension-timber or "tie" '. During the Saxo-Norman overlap the chancel at Chipping Ongar was roofed, and it seems to indicate the arrival of level eaves courses, upon which both wall plates and sole pieces could be laid. The Ongar rafter couples illustrate a curious interpretation of their own structural behaviour, as in extension at and above their collars; but in compression beneath them, and whilst incorporating level eaves courses, they do not provide any evidence for base-tying, but were designed to stand as virtual arches, retained by their imposts.

With the next roof system at Waltham Abbey striking progress had been made, the stages of which cannot as yet be deduced. The roof was designed to be base-tied at every couple, a fact that may indicate a preoccupation with timber ceilings, which would seem probable before the adoption of either groined or ribbed vaults; but the cause and effect relationship betwixt the two is a matter for research. European roofs that were base-tied in like manner seem relatively numerous; St. Pierre de Montmartre, of c. 1147; St. George, Hagenau; and St. Germain des Pres at Paris; the last two of which are ascribed to the 12th century by Deneux (H. Deneux, C.R.M.H., 1926). This roof tends to assist with the evaluation of slight evidence, since a single piece of evidence must be weighted according to the social level of the patronage to which it owes its existence. In the case of Waltham Abbey's roof we have the patronage of the king— possibly Henry I—and the work, although unique at this time, must be considered to be of national significance: if it did not reflect the national trend it probably set it subsequently, and was designed with an awareness of wider Continental developments.

Until the middle years of the 13th century, and particularly until the end of our Norman period, c. 1150, there seems to have been one overall sequence of developments in the roofing of great churches. During the Early English period this divided into several separate series which were contemporary, creating a spectrum for which it is difficult to produce a chronological diagram.

With the oldest surviving high-roof of Wells Cathedral we have a direct development of the Waltham Abbey design, which no longer needed provision for a ceiling, since it was over a ribbed vault; and joined the two collars by a mountant, which was jointed to resist the sagging stresses generated by the soulaces transferring the rafters' spreading moment to the lower collars. In this example the eaves were converted into parapets during the early 14th century, which is unfortunate for the present purpose, but tie-beams were fitted to every *fourth* couple.

It cannot be deduced when or where the scissor-braced category of roofs first appeared in England, but the nave of Peterborough Cathedral is for the present a possibility that has admittedly been inadequately researched, and thereafter the eastern transept roofs of Lincoln Cathedral. The type was, however; established by 1200, and was thereafter widely used until at Salisbury Cathedral its functions were reversed from resisting extension to resisting compression, when long timbers were no longer required. By the same approximate date the series of Lincoln roofs based on truncated secondary rafters had begun and was to terminate in the south main transept there, completed by c. 1320; the same type was to appear above the nave of Beverley Minster during the 13th century. The scissor timbers themselves probably derived from the preceding types with seven-canted archivolts and v-struts above their collars, which visually resemble scissors. The contemporaneous development of scissor braces and secondary- or under rafters is illustrated by the roof surviving at Kersey Priory in Suffolk, from c. 1250 (C. A. Hewett, 1976, 48). By the later 13th century the naves of Chichester Cathedral and Boxgrove Priory had been roofed, Chichester having tall crown posts supported by timbers bearing remarkable visual affinities with secondary rafters. The first jowls also occur in the Chichester roof.

Two major preoccupations of the master carpenters during these years, c. 1180 to c. 1300, were roof-racking and base-tying, the production of pitches that remained flat having already been achieved. The bay lengths of almost all great church roofs remained very short until as late as c. 1250-60, and the earlier examples such as those at Lincoln were provided with tie beams at, on average, every fifth rafter couple. For this feature, too, there are Continental parallels, such as Troyes Cathedral south transept, which has tie beams to every sixth couple (H. Deneux,

321

C.R.M.H., 1926). During the later part of this period even the closely-spaced tie beams provided were improved, and in the eastern roofs of Westminster Abbey and Salisbury Cathedral branch-ended ties were introduced; these were still spaced closely, but effectually tied the wall plates at more frequent intervals.

The racking, or inclination from the vertical, that is inherent in any long series of rafter couples connected only by their external cladding, had by *c.* 1290 prompted at least three anti-rack framing systems, the least effective of which was diagonally-disposed rafter braces (as fitted at the Gloucester Blackfriars church and the eastern roof of Westminster Abbey). Lincoln Cathedral had, it seems, introduced collar purlins, first as single and central members, and thereafter in the Angel Choir as three purlins, all laid over the collars with the lower pair placed close beneath the pitch surfaces. However, the most important step was taken at Exeter Cathedral, probably by *c.* 1290 (J. H. Harvey, 1974, 25): the high-roof of the eastern presbytery had been designed with side purlins that were combined with scissor-braced couples.

From this point developments were numerous and rapid, a fact which suggests that there were better communications than are normally held possible; Salisbury's north-eastern transept was fitted with side purlins mounted upon their own anti-rack arcade, and at Exeter the great Master Thomas Witney had designed his continuation of the high-roof. Witney was, as Dr. Harvey said, 'one of the few proven examples of an architect who was both mason and carpenter, and an outstandingly advanced designer'. In this roof design elements which were discernible as early as the Waltham Abbey roof were combined with almost every development since to produce the magnificent system illustrated in Fig. 135. Tie beams were dispensed with, short bay lengths retained, and racking combatted effectively by both collar- and side purlins; it is also interesting to note the adoption of the high position of the scissor braces, which were another retained feature. In this roof recognisable crown posts appear in a form that seems to derive from the mountants fitted at Wells.

From this time (the roof was completed by *c.* 1342), the insistence on frequent base-tying was phased out, and with the Chichester roof tie beams were already being placed at every ninth couple. A strange lack of interest in king-post design is evident from these examples, despite their early appearance at St. Augustine's Abbey in Canterbury. This (if the fragmentary nature of surviving evidence is at all valid) indicates a departure from European developments, since Notre Dame employed composite king posts with two central purlins between 1230 and 1245, as did the church at St. Leu and St. Gilles, Paris, in 1320.

Another complexity had been added to this aspect of carpentry by the middle of the 13th century: the exploitation of 'compassed' roofs, an early example of which is provided by the Greyfriars' Church at Lincoln, and a late example by the choir high-roof at Carlisle Cathedral. Low-pitched roof designs were certainly advocated as early as *c.*1325 by the unknown master of Bristol Cathedral's choir high-roof, and combinations of all the principles described were probable after this time, including that of built beams, which seem to appear first at Bristol.

Evidence for roof designs during the Decorated period is relatively sparse unless buildings of levels beneath the great churches are included, as they must be, and as far as the south-eastern counties are concerned crown-posted types predominated. However, during the 14th century, designs that were arched to their collars were built in the parish churches almost as frequently as were crown-posted roofs, and this led to several interesting developments. Perhaps the most remarkable was the introduction, not later than *c.* 1350 at St. Clere's Hall, of principal-couples that were arched to their collars and exerted a tying action on their top-plates. This was probably the final stage of removing visual obstructions from the base-levels of timber roofs. According to this evidence base-tying at close intervals had been introduced, and was followed by the development of self-tying rafter couples that were tied sparingly, with beams placed at ever longer bay intervals. In the majority of examples, however, during this process collars had remained common, and were for the most part fitted to every couple; it was at collar level that the next reduction was to be effected.

The church roofs at Margaretting and Cressing in Essex are equal in quality to monastic or cathedral works, and they illustrate the integration of well-tried roof designs into very complex hybrids. Two important aspects of this were the use of collar- and side purlins together, and the retention of the crown post in its original high position, as used at Exeter. Other examples are provided by the barn and the granary at Rookwood Hall, Abbess Roding (not illustrated, *see* Hewett, 1969, 120), and the service-wing roof at Horham Hall, Thaxted. This was to lead to the ultimately predominant use of side purlins, but one significant document of that process survives; the nave roof of the church at Little Braxted. This has side purlins which are wind-braced in the pitch planes, but retains common collars in addition, emphasising by its solitary example the fact that the collar level was the last to be cleared of visual obstructions. This ultimate selection of the side purlin (set in-pitch) as the most suitable form of roof had begun at least as early as in the eastern arm of Exeter, the purlin itself having appeared even earlier in the context of lean-to roofs.

The sequence of events described cannot be said to constitute a progression, other than towards the least complexity that would endure or suffice; and the economic factors of cost and diminishing material resources were controlling factors. Later trends in roofing are also illustrated by the examples; true queen posts were used in the Queen's House in 1528, and again in the great roof of Horham Hall by 1575, and thereafter throughout the 17th century. At Horham these were combined with raking struts, and it is probably significant that at Westley Hall in Suffolk the queen posts were omitted and only raking struts used, a feature which is reflected in many Elizabethan barns in Essex. Westley Hall is dated by carved numerals to within the last decade of the 16th century. During the 18th century suspended king posts with iron foot-straps appear, and a dated example from 1750 exists in the barn at Porter's, Felsted, Essex.

The succession of house types

In south-east England there has been a triple series of developments in this field, all of which derive from the great aisled halls exemplified by the Bury at Clavering; the earliest surviving examples of these had rectangular ground-plans and external walls of a single storey's height. That the origin of this type must pre-date the Conquest is obvious, and the barn at Belchamp St. Paul's illustrates all the necessary structural features of the type, which in the case of Clavering's capital manor was a royal one. A good monastic example survives at Great Cornard in Suffolk (C. A. Hewett, 1976, 53). Essentially this type contained a great hall, open to its ridge, with bays at either end partitioned off for service and solar purposes; and it was in these end-bays, it seems, that lodged floors were fitted during the 13th century. There were no structural differences between the big aisled barns and such aisled halls. As to internal economy, the halls were most frequently two bays in length, and were equipped with smoke exits or fumers (L. F. Salzman, 465), while the service end was divided into two rooms for a buttery and pantry, which necessitated two service doors. Entry was normally by a main door situated against the service end partition wall, and the exit placed opposite, across the hall. Stanton's Hall and Tiptoft's Hall contain rare survivals of three service doors (Fig. 119), the central one giving access to the external kitchen building, an example of which is at Little Braxted (Fig. 187). The two principle variations of this archetype were the H-plan hall houses and the miniature copies created by those who possessed both the freedom and the capital to build them.

Priory Place furnishes an example of the experimental hall house with two jetties, at its service and solar ends, which resulted in an H-plan at first-floor level whilst retaining the rectangular ground-plan. Tiptoft's Hall was more experimental in structural method, and provides evidence for the advocation of the H-plan at a very early date; while Baythorne Hall illustrates the further development of this type, which was achieving structural stability before c. 1300. Among the last examples built with two cross wings, both jettied, and an open central hall, was Jacobe's at Brightlingsea, by which time tastes in living accommodation had already begun to change. Jacobe's therefore represents the end of an architectural tradition that derives from feudalism and manorial law.

During the earlier 15th century small houses were built for priests, and particularly for chantry priests, because the endowment of such chantries constituted a type of insurance policy for the devout, or the repentant. Many of these priestly residences have been identified, particularly in Essex, and the usual type was a single wing, potentially a cross wing, combining the facilities of both the service- and the solar wing of the H-plan hall houses. This was evidently deemed sufficient for the celibate priest who had his principal chamber on its first floor. As in the example shown at Maldon (Figs. 181 and 182) these not infrequently became vicarages or rectories when (as at Maldon) the H-plan was completed by the addition of a hall range with solar cross wing.

Throughout this period, from the early 13th century until the Dissolution, the rectangular ground-plan continued, and early examples are provided by the Prior's Chamber at Prittlewell and Rayne Hall. A modification of the original purpose is illustrated by Paycocke's House, which was storeyed through a length of five bays, and equipped with facilities for the pursuit of Paycocke's trade as a wool merchant. The Prior's Chamber illustrates the fact that from some time close to c. 1300 halls had been elevated to the first-floors; while Jacobe's Hall and St. Clere's Hall demonstrate that opinions were divided as to which design was most appropriate, for both persist with open halls on their ground-floors.

By 1528 the new residence for the Lieutenant of the Tower had been built, and with this example all the domestic characteristics that were to predominate during the 17th century were in evidence. However, this process of change was not simple or direct, and numerous houses are known that afford clear evidence that both carpenter and patron changed the design during the process of the building operation. Some examples of these transitional buildings have been published in detail (C. A. Hewett, 1976, 85, 83 and 77), and what they illustrate is a change to houses designed in two parts, divided by the central brick chimney-stacks that were an architectural feature. Façade gables, such as provided for the Queen's House and the residential range of Staple Inn, often appear on the fronts of these two-part houses; and the stairs towers invented to meet the need for new stairs where an open hall had been laterally divided by a first floor, were frequently retained as architectural features (as is illustrated by Doe's Farmhouse in Fig. 202).

The third succession of house types deriving from the great manorial hall houses were the miniature copies of them, which were built in some numbers as the yeoman class multiplied. The oldest among these is Songers (Fig. 74), and a 14th-century example is provided by the Bridge House at Fyfield, while the house at Deal Tree Farm in Hook End is of a 16th-century date.

Conclusion

The hypothesis that carpenters' joints underwent processes of development towards mechanical efficiency has now been illustrated by the series of timber buildings described. It might have been possible, as an exercise in ineptitude, to select a series that did not establish the developments; but it is unlikely that any series, chosen at random, would have disproved it. It is also apparent from the same series that the different categories of joints did not evolve simultaneously, which makes it unlikely that any timber building will incorporate the most efficient forms of several categories of joints. The scarfing of timbers into long top plates was perfected by the last decade of the 13th century at Place House, Ware, at which time the developments of integral framed floors and jetties were both in their experimental stages. It is evident that the overall systems by which timber buildings were produced also changed, a fact that rendered certain types of perfected joints unnecessary; they lapsed into obsolescence.

During the time of the Romanesque styles, both Saxon and Norman, and the Early English period, the system used for timber building seems to have been constant, using long and continuous passing braces to stabilise frames of considerable size and complexity by sideways assembly after the rearing. This system produced notched-lap-joints, which attained their supreme mechanical efficiency between *c.* 1213 and *c.* 1250, but which were to become unnecessary when bracing became part of the rearing process. By the time the Decorated style was developing the use of chase-tenoning was replacing lap joints, and the majority of brace timbers were designed to act in compression. The reason for this major change of system is a matter of speculation, as is the change of preference that accompanied it from straight to curved timber. That the processes of rearing, and the order of assembling frames, became far more complex as a result of these changes is evident, and is well illustrated by the timber tower of Blackmore church.

The changing tastes in Gothic style did not directly affect the development of the carpenters' craft, and it is interesting to assess the degrees to which such styles were expressed in timber. The two great barns at Cressing were built during the Early English period, for example, and the extent to which they conform to that architectural style, or to the Gothic style, has to be determined. If Sir Banister Fletcher's comparative analysis betwixt the Gothic and the Renaissance styles is consulted (Sir Banister Fletcher, 1956, 601), it will be seen that these barns definitely expressed the style, but only in its most general terms, and inasmuch as they were produced by 'a prevalent and eager national taste', as Ruskin said. After the transition from Early English into the Decorated style a more direct expression of that type of Gothic is discernible; there must be a close relationship between this and the change of preference in timber, which had been for straight timber before curvature was introduced wherever structural soundness permitted it. The end wall of the Fressingfield building may be interpreted as a manifestation of the Curvilinear Decorated style, with its prolific wind-bracing that visually suggests a section through a whole tree of symmetrical growth. The timber buildings are the carpenters' expressions of the national style, and as such show Gothic conceived in terms of framing, a concept with which we are no longer familiar and which was absolutely foreign to the masons, on whose very different achievements the general conceptions of the styles are based.

As has been suggested with regard to mouldings, stone and timber are so dissimilar that more widely differing versions of Gothic might have resulted. With timber carcasing such as was used for both houses and barns, few of the normal features of stone buildings were adapted for use, and it is to the credit of medieval carpenters that they expressed the style almost exclusively in the terms of their own material. The buttress and the arch buttress, together with the pinnacle and the crocket, were not normally translated into timber; the various arcatures peculiar to the styles were reflected, as was the tracery of windows together with the device of the attached shaft. In these last cases (as with the case of framed vaults, which were essentially 'mock' vaults) the carpenters copied from the masons, but the examples of pure carpentry express the styles independently; the Blackmore tower would be Perpendicular without its cinquefoiled windows, as would the front elevation of Paycocke's House.

No attempt has been made to include every variety of scarf joint noted during field studies, because many represent side-shoots that depart from the main course of development, and for that reason did not influence subsequent methods. The same has not been found to be true of joist joints, which points to the lesser flexibility of the physical requirements involved, all floors having to support weights as horizontal planes; while some scarfs were designed to occupy supported situations, such as those for wall plates, and others necessarily occupied 'flying' situations, such as arcade plates.

Whatever the practicable variations of any specific joint, they will have been made by some carpenter at some date. A few such examples have been described, but none warrant inclusion in the overall course of development, and all can be dated by reference to the national historical sequence. Throughout our medieval period the monarch was absolute, and the best craftsmen had little choice of employer, who was almost inevitably the Crown. The great churches themselves, like the monastery of Ely, borrowed the best craftsmen from the king—Master William of Hurle, for example, who had been appointed *capitalis carpentarius* with authority *mutatis mutandis* by Edward III (F. R. Chapman, 1907, 46). Master Hurle, or Hurley, was therefore the carpenter of England; an impressive title. Of other carpenters employed at Ely it has been stated: 'craftsmens' surnames appear in the King's Rolls of the same dates' (F. R. Chapman, 1907). The technological development of carpentry, like that of other crafts, resulted from the royal patronage of the finest craftsmen, and all that is debatable is the period of time necessary for their developments to become available to the majority of craftsmen.

326

Index One

by Grace Holmes

PERSONS AND PLACES

The numbers in italics in this Index relate to Text Figure Numbers

Tymperley, Frances and Roger, 194
Tymperley's, Colchester, 194, 287, 306

U.S.A., 248, 269, 271
Upminster, Essex: 184; Ch., 85, 99, 105, 293

Virtue, Robert and William, 216

Wadhams, M. C., 199
Wakes Colne, Essex, 64, 72, 73, 263, 265, 266, 297
Walden Abbey, Essex, 218
Walker, Samuel, 43
Walker's Manor, Farnham, Essex, 175, 275
Waltham Abbey, Essex, 50-1, 57, 72, 93, 99, 169, 291, 321, *46*, *81*
Wantone family, 140, 195, 320
Wantone's, Wimbish, 126
Ware, Herts: 265, 271, 299, 301, 318, 325; Manor, 122-3
Warkworth Castle, Northumberland, 108
Watney, Daniel, 231
Welles, Joan, Walter and William de, 180
Wells: Cathedral, 2, 63-4, 69, 72, 91-3, 105-6, 136, 167, 215, 241, 292, 297; Dean and Chapter, 167; Chapter House, 136-7, 266, *120*; Chequer, 167, 213, 318, *153*; Lady Chapel, 153-4, 318, *138*; Library, 195, 227, 280, 281, 287, 318, 322; Roofs: arcuation, 101, 322; framed vault, 119; main span, 290, 291; nave, 69-71, 72, 83, 99, 291,

Wells (cont.)
59, 71; n. transept, 241, 260, *216*; n. triforium, 65-6, 72, *55*; retroquire, 153, 158, 170, *137*; tower crossing, 292; Vicars' Close, No. 22, 168, 280, 287, 308, 318, *154*
Wenyngton Manor, Essex, 159
West Bergholt, Essex: Ch., 27, 34, 289, 291, *25*
West Mersea: Ch., 41, 58, *37, 38*
Westley Hall, Suffolk, 323
Westminster Abbey: 25-6, 99, 292; Door, 25-6, *23, 24*; e. apse, 113, *100*; Henry VII Chapel, 169; n. transept roof, 112-3, 170, 215, *99*
Westminster Hall: roof, 170, 188, 203, 319
Weston, Richard, 194
Wethersfield, Essex: Ch., 85, 86, 99, 279, 293, *73*
Wheathampstead, –., (architect), 116
White Roding, Essex: Ch., 97, 170, 263, 273, *85, 246*
Whitham, J. A., 148
Widdington, Essex, 175, 204, 267, 271, 275
Wiggon's Farm, Helion's Bumpstead, 269
William I, king, 46
William, Lord Montagu, 156
William of Hurle, 161, 326
William of Sens, 82
William of Valence, 108, 109
William of Wykeham, 157
William the Englishman, 82
Willingale, Essex, 269
Wilson: C. G., 136; Sir William, 243

Wiltshire *see* Calne, Longleat, Malmesbury, Salisbury
Wimbish, Essex, 126-8, 140, 171, 264, 279, 286, 293, 299, 320
Winchester: Castle, Round Table, 279, 286; Cathedral: 157; choir stalls, 136, 141, 172, 293, 301, 318; choirboys' vestry, 297; nave roof, 80, 93, 157, 176, 246, 260, 264, 271, *142, 221*; n. transept, 214; presbytery roof, 176, 204, *158*; *see also* St. Cross
Winchester, Countess of (13th cent.), 123
Wingfield College, Suffolk, 265
Witham, Essex: 59, 200-1, 281, 287, 306-7; Newland Street, 317
Witney, Thomas, 149, 291, 322
Wood, R. G. E., 108
Worcester Cathedral, 56, 64, 244, 260, 297, *220*
Wren, Sir Christopher, 204, 241, 260
Wyford, William, 157
Wykehampton, Richard de, 110
Wynter, Alan, 110
Wynter's Armourie, Magdalen Laver, 72, 110-11, 172, 199, 266, 293, *97, 98*

Yeavering, Northumberland, 72
Yevele, Henry, 183
York Minster: Chapter House spire, 113, 117-20, 161, *104, 105, 106, 107*; nave and nave aisles roof, 146, 169, *130*; n. transept triforium roof, 181, 204, *165*
Yorkshire *see* Selby

Index Two

by Grace Holmes

SUBJECTS AND TERMS

Abutments: bird's mouthed, 268, 271; bridled, 170, 174, 270-1, 305, 312, 318, edge-halved, 305, 312; feather-wedged, 110; over-squinted, 268, 271; sallied bridling, 271; scribed, 210, 227, 228, 287, 292; square 174, 263, 264, 269, 270, 271; squinted, 269, 270; tapered, 267; terminal, 270, 292; under-squinted, 110, 115, 264-5, 268, 270, 271; vertical, 174, 267
aisled buildings, 23-4, 32, 33, 40, 43-5, 47, 53-5, 59, 64, 72, 100, 159-60, 171, 173-4, 184, 276, 323

angles, refined entry, 109, 289, 291, 292
Anglo-Saxon: carpentry, 5-34, 56, 57, 276, 320-1, 325; floors, 202; towers, 202
Anglo-Saxon Chronicle, 31
apses, 35, 113
arcades, 55, 56, 63, 73, 78, 101, 303
arcature, 21, 325
arches: 3, 35; collar, 188, 191, 197, 239, 242; inverted, 167
see also cruck building
architecture, English, 1, 2, 73
see also Anglo-Saxon, Decorated, Early English, Gothic, Perpendicular, Tudor

archivolts, 110, 134, 166, 274, 301, 321
arcuation, 101, 275, 322
arris, arris trenches, 15, 16, 21, 93, 126, 142, 146, 258
ashlar: 50, 57, 69, 72, 75, 89, 110, 139, 148, 158, 166, 182, 241, 246, 254, 260; cants, 35, 112; plate, 97
assembly: 1, 162, 275-6, 325; main span, 184, 276, mainspan reversed, 276; platform, 85, 162; reversed, 33, 72, 88, 174-5, 184, 276; scribed, 281; triple-centred, 189
astragal, 55